Mildred Cooke

Robert Cecil 1st Earl of Salisbury

zabeth Cope   The Marquesses of Salisbury, of Hatfield House

dy Elizabeth Egerton

dy Frances Manners

dy Anne Cavendish

zabeth Brownlow

nnah-Sophia Chambers

arlotte Gornier

ah Hoggins ('The Cottage Countess')

bella Poyntz

dy Georgina Pakenham

bella Whichcote

n. Myra Orde-Powlett

ith de Telegd(1) = Martin, 7th Marquess of Exeter (1909-1988) = (2) Lillian Johnson

Michael, 8th Marquess of Exeter (b.1935) = Nancy Meeker   Lady Marina Castonguay

Anthony, Lord Burghley (b.1970)   Lady Angela Cecil (b.1975)

ONE HEART, ONE WAY

THE LIFE AND
LEGACY OF
MARTIN EXETER

CHRIS FOSTER

FOUNDATION HOUSE PUBLICATIONS
Denver, Colorado   Vancouver, British Columbia   London, England

© 1989 by Foundation House Publications, Inc.
All rights reserved. No part of this book may be reproduced or transmitted in any form without permission in writing from the publisher, except in the case of brief quotations embodied in articles and reviews.

Canadian Cataloguing in Publication Data
Foster, Chris, 1932–
   One heart, one way : the life and legacy of Martin Exeter
   ISBN 0-921790-00-7
     1. Exeter, Martin, 1909–1988. 2. Pioneers—British Columbia—Biography. 3. Emissaries of Divine Light—Biography. I. Title.
   BP605.E4F68 1989   289.9   C88-090435-6

British Library Cataloguing in Publication Data
Foster, Christopher Joseph, *1932*–
   One heart, one way: the life and legacy of Martin Exeter.
   1. Great Britain. Exeter, William Martin Alleyne Cecil, Marquis of, 1909–1988 I. Title
   941.082′092′4
   ISBN 0-921790-00-7

Lines from *The Healing of The Planet Earth,* by Alan Cohen, copyright © 1987, reprinted with permission of Alan Cohen Publications, P.O. Box 5658, Somerset, NJ 08875, U.S.A. Lines from *In Search of Identity,* by Anwar el-Sadat, copyright © 1977, 1978, reprinted with permission of Harper and Row, Publishers. Permission in the U.K. granted by William Collins Sons and Co. Ltd. Lines from *A Woman of Egypt,* by Jehan Sadat, copyright © 1987, reprinted with permission of Simon and Schuster.

Inquiries, orders, or catalogue requests should be addressed to:
    Foundation House Publications, Inc.
    4817 N. County Rd. 29
    Loveland, Colorado   80538
    USA

Foundation House Publications (Canada)
P.O. Box 9
100 Mile House, B.C.
Canada
V0K 2E0

Foundation House Publications (U.K.)
Mickleton House
Mickleton
Gloucestershire GL55 6RY
England

Printed in Canada.
Designed by Janice Wheeler.

*Never underestimate the power
of spiritual expression.*
MARTIN EXETER

CONTENTS

FOREWORD 11
ACKNOWLEDGMENTS 14
AUTHOR'S NOTE 15

1.
A CALL TO GREATNESS 17

2.
MEETING MARTIN 22

3.
AN ILLUSTRIOUS HERITAGE 25

4.
EARLY YEARS 32

5.
THE ABILITY TO OBEY 42

6.
INTO THE UNKNOWN 52

7.
"SIMPLIFY, SIMPLIFY" 57

8.
COWBOY 65

9.
LIFE UNFOLDING 76

10.
A FULL PLATE 83

11.
LLOYD MEEKER 93

12.
TOUCHSTONE 101

13.
TURNING POINT 107

14.
SPIRITUAL PIONEERING 117

15.
A LIGHT IN THE EARTH 124

16.
A CHALLENGE ACCEPTED 137

17.
DEMANDING YEARS 145

18.
A MAGNIFICENT WHOLE   156

19.
A DECADE OF CHANGE   173

20.
LETTERS FOR LIVING   191

21.
EAST MEETS WEST   202

22.
THE TRUE STEWARD   214

23.
THE LIVING PROOF   226

24.
A QUANTUM LEAP   233

25.
A HALO AROUND THE EARTH   241

26.
OF STARS AND SOLAR SYSTEMS   253

27.
THE RISING TIDE   260

28.
FINISHING THE WORK   267

29.
OUR TRUE ANCESTRY   280

FOREWORD

I was a day older than Martin when we joined Dartmouth Royal Naval College together in 1922 at the age of thirteen, and that was a valuable asset in a friendship which continued throughout our long lives. We had known each other before that as neighbours in the school holidays, but arriving at Dartmouth an instinctive bond drew us together that was never lost, although our interests and natural aptitudes were very different and our lives far apart later on.

I played games of every kind and was good at some of them. But it was only athletics and long distance running and fitness that mattered for Martin, largely to follow his older brother David Burghley, world champion hurdler at that time and later, whom he admired greatly. I enjoyed mathematics and other subjects but was hopeless at the detailed drawings, carpentry, and engineering in which Martin excelled. Yet we were always together in our free time—the two of us would go off, often in silence, for hours. Looking back, I think it was that need for silence which drew us and kept us together. It was a need, though we certainly didn't realize it then, for an inner communion, which became for Martin an essential and natural part of life, though it is still rare in the educational systems of today.

However the firm basis which was provided for us in the standards of the day—strengthened in our subsequent life together at sea—was important in all that Martin would accomplish later on. Naval discipline combined with a constant need for responsible and independent thinking, and leadership in all kinds of different circumstances, prepared the way for the remarkable achievement which Chris Foster recounts so well.

To acquire the necessary skills at the age of twenty-one to build the new 100 Mile Lodge and then develop the cattle ranch into the large and prosperous enterprise which I came to visit a few years later was a formidable undertaking. The decision that came next, not to rejoin the Navy on the declaration of war, must have been the most difficult of all for Martin, since his new understanding of the true spiritual life was only beginning to take shape firmly in his mind. From my own standpoint, still in the Navy, it was totally right, both in considering the need for food production and also the far more important need to build up his emerging spiritual life, which had to be nurtured carefully. However, I remember the difficulty of trying to convince his family of this, particularly his father and mother, imbued as they were with the long family tradition of service and duty to the nation. In time they came to accept it, and to their great credit were never even critical afterwards, though I doubt if they ever fully realized the importance of the great spiritual work their son was undertaking.

From the time of Martin's first meeting with Lloyd Meeker there was never any doubt in his mind what should be done, only a certainty of what was right and that the doing of it would bring its own fulfilment, as indeed it has abundantly. Whenever Martin came over to England, whether to Burghley House or elsewhere, we would meet and talk together of so much that mattered deeply to both of us. But for Martin talk was of little value. It was the living and doing that mattered. My path had changed a long way too from the one we shared in the first place, but it had moved in a rather different direction, becoming more influenced by the teachings of the East; I think it saddened him that I could not share wholly with him in his life.

The great thing for both of us though and for all those who read this account is the acceptance of the spiritual life as Reality. It is no longer a matter of faith or belief. It is real. Martin has shown us the reality of that spiritual life, and how we too can develop it and live in a highly efficient, practical way at the same time.

From the small beginnings of Sunrise Ranch and 100 Mile House to the present worldwide scope of the Emissary organization has been a very large step. Now, in the way that destiny seems to order for our good, the further development of that work falls naturally to Martin's son, Michael, now 8th Marquess of Exeter, whose godfather I am proud to be, and to his wife Nancy. The work is in safe hands and I wish them well. This book tells us beautifully how it all came about.

      Michael Culme-Seymour
      Powerstock, Dorset

ACKNOWLEDGMENTS

My special thanks to all those who helped bring this book to completion, including Richard Baltzell, Richard Heinberg, Brian Scrivener, Nancy Exeter, Judy Smookler, and Joy Foster, each of whom provided valuable editorial assistance, and Norman Smookler, my publisher, always a cheery light along the way. A warm thank-you also to Janice Wheeler, who designed the cover and the interior.

AUTHOR'S NOTE

Many more people played a part in the unfoldment of Martin Exeter's life and purpose than I could possibly mention in this book. This is one person's record—one person's perspective of this remarkable man and his legacy.

I.

A CALL TO GREATNESS

In the serene and spacious assembly room at 100 Mile House, British Columbia, where Martin Exeter gave so freely of the fruits of his living, his wisdom, and his uncompromising integrity, some three hundred people gathered to remember him. Several thousand more in twenty countries around the world, from Israel to Japan, from Argentina to Australia, shared through teleconference connections.

Sadness attended the event, of course, but it was a sweet sadness, overshadowed or even transformed by a deep thankfulness for what this man had offered during his life, and by a common determination to continue his work.

He took a few cubic feet of fairly average human flesh— flesh imbued initially with the same fears and confusion as anyone else's—and shone through it the unwavering light of a stable, true, and loving spirit. Not everyone who knew of him or came into contact with him understood what he was doing. Most were so engrossed in their pursuits that they were not even interested in understanding. A few, however, myself included, drew closer to see this remarkable sight: a man who refused to blame or judge anyone else for what happened in his world, but simply took responsibility to express that steady, loving spirit in every moment.

These few not only drew closer. They aspired to let the same changes and transformation happen in themselves, to let the same necessary discipline rule their own behavior. Thus there spread across the planet the beginnings of a "new nation," a family of men and women bound not by rituals or systems of belief but by the practical expression of a single unifying spirit, the universal spirit of love and truth.

Martin's son, Michael, opened the memorial service. He had succeeded to his father's title, becoming the 8th Marquess of Exeter; he had also inherited the task of coordinating this "new nation"—the Emissaries, as they are named. Lean and erect, his passion burning safely like a flame enclosed, Michael pointed out that his father bore a noble name in the hereditary sense, but it was his noble spirit that touched the hearts of many and was being honored now. It has been said that a person is fortunate to gain half-a-dozen genuine friends during his or her life. "In this instance," Michael remarked, "I would wager that there are thousands who count their connection with Martin to be intimate and tender."

Ross Marks, former mayor of the village of 100 Mile House and a close friend of Martin for forty years, spoke of the qualities of character that he revealed in his life, forged and tempered when he came to Canada as a young man in 1930 to manage his father's cattle ranch. "He had a hands-on experience," said Marks. "Very hands-on. I don't think anyone who lived in the Cariboo in those Depression days just sat back and watched the universe unfold! He was a cowboy, a plumber, an electrician, a carpenter—one could go on down the list—essentially untrained, but with the ability to see something in front of him that needed doing and do it. In my view that was one of his outstanding traits—it had some rub-off effect on me, I'm glad to say." An ebullient man with a keen sense of humor himself, Marks spoke of Martin's booming laugh. "Very often when he told a joke he laughed first, and that inspired others. He always appreciated humor and saw it as a way of loosening things up."

Speaking from the Emissary international headquarters at Sunrise Ranch, Colorado, Roger de Winton told how Martin directed courses in the art of living for many years, helping people from a wide variety of backgrounds and cultures to develop a more effective approach toward life. Jim Wellemeyer, who lives at the Green Pastures community in New England, summed up Martin's philosophy: "Things really aren't as they seem. Regardless of what is appearing in this external world, there is that which is of far more value present with us."

Lastly the bell-clear, musical voice of Rupert Maskell resounded over the telephone line. He spoke from the Hohenort Hotel in Cape Town, South Africa, one of twelve primary Emissary communities, or "Units," as they are called, that girdle the globe. As he intimated, the scene at the Hohenort was very different from that in 100 Mile House. Instead of a cold winter's afternoon it was after midnight on an African summer night, with crickets singing outside and stars twinkling brightly overhead. Yet the universal spirit was just as much present there as it was in the Cariboo. Maskell told how he had spoken earlier in the evening with a man named Absalom Makhathini in Zululand. "Although he is unable to share in this teleconference this evening he asked me to present his special wishes and those of the people with him at Empumalanga," Maskell said. "I would just like to say two words of Zulu to represent Absalom here: Ngiyabonga Nkulunkulu—'Thankyou, Lord.' Thank you for the life that was revealed so beautifully in Martin, and thank you for the universality of it which allows us to experience it in its fullness right now."

Often, when a leader with spiritual vision passes on, the real message that he or she brought is lost. Followers fall away, or conflict emerges. This happens inevitably if those concerned have merely basked in the light shone by another. In this instance this was not the case. Martin Exeter had done his job well. Quietly, tenderly, but firmly, he insisted that those who loved him take responsibility to reveal the same quality of spirit that he revealed: thus they were in position to continue strongly without him.

The way which this man discovered and articulated for others is not a new way. Wise ones of all ages have spoken of it, and walked in it. It is a way of adventure, nobility, and common sense—a way that honors the importance of spirit in every little circumstance and moment of our living. We are spiritual beings and we ignore that fact at our peril. The problems that confront us, individually and collectively, cannot be satisfactorily solved by grappling with them, because in the final analysis they simply relate to an absence of the true qualities of spirit. Once those qualities are allowed release, as if by magic the whole unhappy picture changes.

Humankind is the vehicle created by spirit to make possible the action of spirit on earth. William Martin Alleyne Cecil, as he moved through the many roles and exigencies of his life, from British aristocrat and junior naval officer to cowboy, founder of a village, and spiritual mentor, came to a vivid realization of this central truth, and of his own absolute responsibility to reveal that truth in his living.

A person who does this carries considerable force, regardless of his or her position in the external scheme of things. Just as a mountain has power to evoke a sense of profundity and stature by its mere presence, so was it with Martin Exeter. When he entered a room, everything changed, even though he would probably say very little. His presence was sufficient; indeed, it was everything.

We all have the capacity to admit the power of spirit into our lives, to be the means whereby that power impacts and transforms our world—first, our own miniature world, then the larger one beyond. That is the meaning of Martin's life—the reason I have written this book. Human religions and schools of thought are sometimes exclusive in their approach, erecting fences, saying what must be believed or not believed, and who is acceptable or not acceptable. But the spirit of God is universal, freely available to be known by anyone and to bring comfort and blessing to anyone, whatever their race, color, or creed. There is but one requirement. We must love that radiant center—the spirit—and turn to that. The spirit will then flood us as the sun floods the earth at dawn.

2.

MEETING MARTIN

I first met Martin in August 1956.
I had come out to Canada from England the previous year in response to an undeniable compulsion from within myself; although I had a good job as a reporter on a London daily newspaper, and a promising career ahead of me, an inner voice kept insisting, "There is more to life." Perhaps Walt Whitman was partly to blame. I had carried him with me in my jacket pocket for a year or more, breathing in his seditious thoughts about the nobility and expansiveness of life. But it wasn't just Whitman. I sensed a call, emanating from I knew not where, to rise up out of that familiar routine of "going to work to earn the money to buy the food to get the strength to go to work ... " Walking in the country one day—in the Sussex downs, where white chalk cliffs meet the sea, and skylarks soar and sing—that

secret invitation burst through a last remaining barricade in my heart and I knew that there was nothing else to do but to follow that impulse of spirit wherever it might lead; I remember how I exulted as I agreed within myself to do this.

For some reason—partly it was the thought of forests and mountains and coastal fjords—Canada's westernmost province of British Columbia appealed to me, and I ended up finding a job with the Daily Colonist in Victoria, B.C.'s capital city. Some time later I met Richard Thompson, who with his wife Dorothy represented the Emissary program in Vancouver at that time. I liked Thompson, and wanted to find out more about his philosophy. A few months went by, until one day he invited me to meet Martin during a visit Martin was making to Vancouver.

As I knocked upon the gaily-painted yellow door of the Thompsons' home and stepped inside I sensed that the meeting was an important event in my life. I also sensed that I would know immediately if Martin was genuine. I didn't know how I would know, but I knew that I *would* know. And so it was. I entered the upstairs living room, with its panoramic view of the harbor and mountains, and seagulls flying by, and I knew, even as the man I had come to see rose to his feet to greet me, that I could trust him.

Later I realized what this feeling of trust was based upon. It did not stem from anything that Martin said at that initial meeting: it was based in a sensing of something that in a way was intangible. Here was a man who expressed a quality of character, or of love, unclouded by customary human motivations such as fear, resentment, and greed. I am not saying such things never rose up in him. Of course they did. He refused to allow them to influence his behavior, that was all.

Such a quality of expression is not always comfortable to

be around. It can, and often did, evoke severe feelings of discomfort, in myself and in others close to him. Sometimes, for example, when Martin gave a talk, the awareness would strike home with an intensity akin to fire that one's own behavior was not yet as clear, or upright, or noble, as inherently one knew it could be. But if love is sometimes a fiery flame, with the power to burn away the ignoble in oneself, it has other aspects too. Its touch can be incredibly gentle—or strong and courageous. Martin revealed all these qualities.

People hold peculiar ideas about love. Many think of it merely as an emotion that surfaces when a man and woman become passionately involved with each other. Some have a more spiritual concept which sees love as a warm, soft, celestial glow: Jesus is often pictured enveloped in this radiance—gentle and meek and kind. But I wager that Jesus was uncomfortable to be around too, much of the time. Love is precise. Love is firm. Love is characterized by truth. Above all, though, love is trustworthy.

Martin demonstrated the true nature of this one great spirit that transcends cultural and geographical boundaries. He also showed—through words, but most of all through his own living—how readily possible it is to find alignment with that spirit so that it becomes one's own.

3.

AN ILLUSTRIOUS HERITAGE

Many people helped to prepare the way so that Martin Exeter could play the role he did in bringing a reminder of man's true nature and purpose to the awareness of humankind. One of these was his famous ancestor, William Cecil, who as Lord High Treasurer and chief adviser to Queen Elizabeth I for many years, helped build England into a dominant world power during the Elizabethan era. This laid the foundations of the British Empire—of which America, of course, was once a part—which in turn spread the English language and culture throughout the world.

In reading accounts of William Cecil's life, distinct parallels emerge between the Lord Treasurer and his descendant, Martin. Both were shrewd, sagacious men. Just as Martin was scrupulous, precise, and careful in his ways, abhorring waste, so was the Lord Treasurer. Just as Martin had the ability to focus on minute details and yet encompass the most sweeping vision, so did William Cecil. But beyond these similarities, the two men were born leaders who because they would not or could not deny their own integrity, inevitably challenged traditional attitudes, shattering the boundaries of the status quo. This kind of impact does not just happen: if they had been "nice" people, concerned to fit in and to be liked, they would probably have lacked the fortitude and stamina to do what was necessary. "What needs to be done?" Martin guided his whole life by that simple question. I suspect the Lord Treasurer did too.

One of William Cecil's primary concerns was to protect and nurture Protestantism, still a vulnerable new offshoot in the early years of Elizabeth's reign. Initiated by Henry VIII, the birth of the Church of England in the sixteenth century brought new life to Christianity, loosening the rigid approach of the Roman Catholic Church at that time. Four centuries later Martin would become involved in a similar process. Ironically, on this occasion it was the Church of England itself—in his experience—which had become restrictive, bound by outdated customs and beliefs. As Martin grew to manhood, a force blossomed in him which did not deny the basic truths of Christianity or any other religion but honored the urgent need for new vision and understanding if the challenges of life were to be met effectively.

People sometimes accused Martin—as they did the Lord Treasurer—of being hard and unfeeling. Such was never really the case. He was one of the kindest men it would be possible to meet. But he was also absolute in his love for the genuine expression of character. That, for Martin, was everything. No worthwhile change would occur in the world without it. He welcomed everyone, being, as the Bhagavad Gita puts it, "of equal grace" to all, and would take immense pains to encourage those who sought understanding. But ultimately his endeavor had but one purpose: that whoever was willing might come to love the spirit of God and the character of God even as he had done. Where there was such willingness one could not help but love this man. Where there was not, sooner or later—probably sooner—a person would tend to feel uncomfortable in his presence, and, as is the human habit, might even blame that discomfort upon him. And so Martin was called many things in his life, not all of them complimentary.

The Cecil family traces its history back to a Robert Sitsilt, who fought with Robert Fitzhamon in the conquest of Glamorganshire in 1091, and later acquired estates in Herefordshire and Monmouthshire. His son, Sir James Sitsilt (or Seisel) died in a siege at Wallingford Castle in 1140. In the latter half of the sixteenth century David Cecil, a descendant of Sir James, trekked south to London as a young man to seek his fortune—and, by Henry, he found it. He caught the eye of Henry VII, being appointed a sergeant-at-arms and enjoying the many privileges and opportunities that came with royal recognition. He fought at Bosworth Field under Henry VIII and had amassed considerable wealth by the time he settled in the small town of Stamford, in Lincolnshire. David Cecil was the grandfather of William Cecil.

England faced a bleak and perilous future when Elizabeth succeeded to the throne in 1558. Wracked internally by fierce feuds between Catholics and Protestants, the nation had become, through the marriage of Phillip of Spain and Mary I, a virtual satellite of Spain. Her wealth and military resources had eroded steadily since the days of Henry VIII. "I never saw England weaker in strength, men, money, and riches," wrote Sir Thomas Smith, a distinguished Cambridge scholar who served under both Edward VI and Elizabeth. With a population of barely four million, the country was at the mercy of the great Catholic powers of Europe—France and Spain. To France especially, which had captured Calais from the English earlier that year and already had a foothold in Scotland, England was a fruit ripe for plucking.

But before long the tide turned dramatically. Elizabeth and William Cecil restored order and prosperity at home, and with some help from the elements, the Royal Navy defeated the proud galleons of the Spanish Armada.

The partnership which made all this possible began when Elizabeth hired Cecil as a surveyor, at twenty pounds a year, to supervise her accounts. He so impressed her with his shrewd, balanced approach that on the morning of her accession to the throne, she appointed him her Principal Secretary of State; it was her first executive step. "This judgment I have of you," she said, "that you will not be corrupted by any manner of gift and that you will be faithful to the state; and that without respect of my private will you will give me that counsel which you think best . . . " Her trust and prescience were well placed. Cecil, whom even Ambassador de Feria of Spain called "a prudent and virtuous man, albeit a heretic," became Lord High Treasurer and Her Majesty's faithful yet strong-minded Chief Counsellor.

Elizabeth and Cecil faced their first stiff challenge in the winter of 1559-60, when Protestants in Scotland rebelled against French domination of their country. The uprising posed a dilemma. If England did nothing, France could move massive reinforcements to Scotland and threaten England's northern frontiers. But if Elizabeth assisted the Protestant leaders, she would risk war with France at an inopportune time. She would also be assisting a rebellion of Scottish people against their legitimate sovereign, Mary Queen of Scots, which troubled Elizabeth's strong moral sense.

Cecil submitted a long memo to his Queen, concluding that she should back a pro-English faction, expel the French, and abolish "idolatry." Elizabeth accepted the bold counsel, although as the crisis deepened, it took a threat of resignation on Cecil's part to persuade her to sanction direct military intervention. The English forces suffered initial setbacks when they attacked the French garrison at Leith, but eventually the fighting swung in England's favor. Cecil traveled north and negotiated a French withdrawal with great skill—helped by the fact that he had cracked their secret code. It was a major triumph. He was made a peer, and became the 1st Lord Burghley.

It was said of William Cecil that, "Though not remarkably tall nor eminently handsome, his person was always agreeable, and became more and more so as he grew in years." His motto, Cor Unum Via Una—One Heart, One Way—in many ways portrayed his own character, for as far as history records he was a man of utmost probity. He guarded the public interest with a zeal that often exasperated his friends and, according to competent judges, profited little in a personal sense from his high office. His attitude toward his enemies was equally singular. He would go to great lengths to effect a reconciliation—so much so that some contemporaries considered him a better enemy than friend!

The Queen deliberately encouraged two or three of her chief ministers to build houses spacious enough to accommodate her court—places where government could be carried on in an emergency. In accordance with the royal wish, Cecil built a stately palace named Burghley House on property he had inherited near Stamford, as well as another large mansion, Theobalds (long since demolished), closer to London. He apparently designed Burghley House himself. It is a huge structure, built over a period of thirty-two years from a local stone so hard and durable that many of the blocks still show masons' identifying marks to this day.

As a boy, Cecil had gone to grammar schools at Grantham and Stamford, and then attended St. John's College, Cambridge. While he was there he fell in love with Mary Cheke, a poor widow's daughter and sister of John Cheke, a noted scholar and Professor of Greek. Against the wishes of his father, who no doubt had something better in mind for him, he married Mary, showing that he was not quite the cold, unfeeling person that some of his critics have suggested.

The Lord High Treasurer died a quiet, peaceful death on the night of August 4, 1598, after calling his family around him and praying for his Queen. Elizabeth broke into tears at the news and never ceased to deplore the loss of her trusted friend and adviser. After an elaborate funeral at Westminster Abbey, the body was carried in state to Stamford and buried in St. Martin's Church, where William Cecil, the 1st Lord Burghley, lies in effigy upon a tomb of colored marble, his wand of office in his hand.

William Cecil's elder son, Thomas, was created 1st Earl of Exeter, initiating the Exeter branch of the Cecil family, while a younger son, Robert, became Earl of Salisbury and founded another dynasty of Cecils at Hatfield House, in Hertfordshire. In 1801 Henry, the 10th Earl of Exeter, was raised to the title of 1st Marquess of Exeter. Through the years following the Elizabethan era, the two branches of the Cecil family played an integral part in many facets of British life.

So when Martin was born in a four-poster bed at Burghley House on April 27, 1909, a rich heritage awaited him in a family long accustomed to wealth and privilege but possessing also a sense of responsibility for the larger world. He joined a brother, David, four, and a sister, Winifred, who was six, in the adventure of childhood.

4.

EARLY YEARS

Surrounded by verdant parkland dotted with trees and sheep—beyond that the twenty-seven-thousand-acre Burghley estate, with its farms and villages—Burghley House in 1909 was a world within itself, a well-ordered community in which everyone had a place and, more importantly, felt a sense of belonging. The First World War had not yet torn apart the traditional class structure of Europe. Old-fashioned virtues such as obedience, courtesy, and respect flowed naturally between those who composed this diverse fabric—from Mr. Pepper, the head keeper, to Westley, the butler; from the warmhearted Mrs. Needs, Martin's nanny, to His Lordship himself. Yes, Lady Exeter kept a close eye on her maids, making sure that they came home by a respectable hour at night. But it was because she felt responsible for them. When a young friend of Martin's gave a tip to Westley that he couldn't really afford, the butler quietly gave it back to the boy as he was leaving. "I think you need this more than I do, sir," he said kindly.

At the center of this world in microcosm stood William, Martin's father. He had become the 5th Marquess of Exeter in 1898 when he was only twenty-two-years old, marrying his wife, Myra, three years later. A conscientious, capable man, William Exeter managed the Burghley estate until his death in 1956, a long, difficult half-century that included two world wars and periods of prolonged agricultural depression. He achieved a considerable reputation in public service during those years, among other things serving as Mayor of Stamford, Lord Lieutenant of Northamptonshire, and, for eleven years, as aide-de-camp to King George V. The latter made him a Knight of the Garter—highest order of British knighthood—in recognition of his contributions to the public good. Despite his formidable mustache and manner, William was a kindly man whose bark—as his children and grandchildren came to realize —was worse than his bite.

Martin did not see a great deal of his father in his early years. In 1914, when Martin was five, William Exeter left to serve as an artillery officer in the Great War, and by the time that was over Martin was attending boarding school and Dartmouth Royal Naval College. A close friendship developed between them in later years, however.

Deep and divisive questions troubled England as Martin passed through his childhood. The Boer War in South Africa had shaken the country's belief in the invincibility of her armed forces—more than that, had challenged the whole concept of imperialism itself. The class system was under fire as more and more people became aware of the poverty that festered in the midst of plenty. One percent of the adult population controlled two-thirds of the nation's capital. A lawyer or doctor might earn up to five thousand pounds a year, an agricultural laborer or a woman in regular employment as little as thirty or forty pounds a year. It was no surprise that a Liberal government had come to power in 1906 with a large majority, and was winning wide support for its program of social reform. In January 1909 the first Old Age Pension Law was instituted, and in the same year Prime Minister Lloyd George brought in his "People's Budget," an endeavor to shift the tax burden to the wealthy. The budget sent shock waves through the House of Lords and provoked an ongoing constitutional crisis when the Conservative majority in the Lords used its veto to kill it.

Women's suffrage was another big issue—Emmeline Pankhurst and her followers staged a hunger strike in 1909 and were to become even more militant. Trade unionism was on the increase and, as always, there was the vexing Irish question. Back of all these things, the new instruments of mass communication, such as radio and telephone and Alfred Harmsworth's halfpenny *Daily Mail,* were having their influence and helping to create a burgeoning middle class.

The major European powers had been busily carving up the world between themselves for nearly a century, jostling and snapping at each other as they did so. When war finally did erupt in August 1914, Myra Exeter accepted the challenge of running Burghley House and raising the children. She had always kept a close, enfolding hand on Winifred, David, and Martin, even though her relationship with them was of the rather formal nature usual at that level of society. At teatime, for instance, properly washed and dressed, the children would be brought down to the drawing room to spend some time with their parents, but their day-to-day care was in the hands of their nanny. Not surprisingly, Martin's love for his mother was characterized by a certain awe. He saw her as a very important person indeed, and once referred to her as "Queen Mummy," because a painting of her at Burghley House showed her dressed in coronation robes with a coronet on her head.

Mrs. Needs, Martin's nanny, was a cheerful, kindly woman from Devonshire who loved children and was incapable of a nasty thought toward anyone. Perhaps something of her nature rubbed off on Martin, for he too grew up with a natural tendency to like people and accept them.

Martin faced his first big responsibility in life at the age of three-and-a-half. Mrs. Needs dressed him in a shiny pink satin suit and he performed as a page at the wedding of Tommy Cecil, a son of his great-uncle, Lord William Cecil.

Some weeks later Martin watched excitedly as an airplane landed in the top park at Burghley. His mother, who kept a thick, leatherbound diary of his childhood years, noted that the pilot, Mr. Hucks, "flew beautifully." On another occasion, while Mrs. Needs was on holiday in her native Devonshire, Lady Exeter helped Martin into a chair saddle for his first pony ride. "He was very solemn," she recorded, "but seemed to enjoy it." Actually, he was nervous of ponies. They were strong, independent-minded creatures who sometimes accelerated into the distance without much thought for a little person sitting on their back. Martin found a way of handling the situation though. If it looked as if things were getting out of hand he simply threw himself off and let the pony carry on without him. He learned to execute this maneuver with a fine finesse so that he suffered the least possible physical pain.

One of Martin's greatest loves—that continued throughout his life—was drawing. His mother saved unused half-sheets from her correspondence for him to draw on. Here, as in many other ways, he was learning by "osmosis." His parents were careful with what they had, and so he grew up being careful too. He never felt a lack where money was concerned, but he was not allowed to feel that he was rich, either; with privilege came responsibility; he had what he needed, no more, and it was to be used properly. One of the toys the Exeter children prized most was a Noah's Ark set, complete with wooden animals, that was brought out on Sundays for them to play with. Because of the careful way it was handled, it lasted for many years, being passed down from one generation to the next.

Martin also loved birds and sent his parents into howls of laughter one Christmas when he asked for "any kind of medium-sized bird (alive)." All the children were encouraged to take care of animals. Martin owned a Belgian hare called Bluejacket, and later, a rough-haired Norwich terrier which he named Nutkin. He would always take Nutkin along when he visited his grandparents at Bolton Hall, in the Yorkshire moors. This was a fun place. There was a pigsty with rats in it where Martin could turn Nutkin loose and watch the chase. There were hills and moors where he and the other children could play and watch their elders shoot grouse. Bluebells carpeted the woods in spring, and platoons of rabbits kept Nutkin forever on the run.

The first son of an aristocratic British family—or any other family—does have some rather unfair advantages. He is likely to be bigger than his younger brother, for a while anyway, and more developed and capable in other ways too.

Certainly this was the case insofar as Martin and his older brother David were concerned. By the time he entered his young manhood David —known formally as Lord Burghley—was a great success socially, with his buoyant, outgoing personality, and was also well on his way to achieving international fame as an athlete, fame which culminated when he won a gold medal in the Olympic 400-meter hurdles at Amsterdam in 1928.

The consequence of this ascendancy on David's part was that from his early years Martin always felt in his brother's shadow, as if the gauge for his own progress was how well he did in comparison to David. He liked his brother, and admired him. But this undercurrent of competitiveness, a feeling on Martin's part that he somehow had to push himself to catch up, was present also. On the other side of the coin, David tended to assume, like any older brother, that he knew more than Martin and that his opinion on any matter was bound to be correct. No doubt this was often the case. But as Martin realized somewhere along the way, it was not always so. Once the boys got into a fierce argument about which direction the guns of a British battleship faced. Martin, who had recently entered Dartmouth, knew how the turrets were aligned—that the fore turrets faced forward, and the rear turrets backward. He could not convince his brother, however, who insisted that a British battleship's guns would always face forward!

In another episode, David and Martin were sharing a room when David became irked at the precise, speedy way in which his brother dressed every morning. No matter what techniques David employed, or how hard he tried—and he tried very hard indeed—he just could not dress as quickly as

Martin. Finally David hit upon a way to get the advantage. With a great shout of glee, he leaped out of bed one morning fully clothed.

Not so humorous—it was a bitter pill to swallow—was the time when Martin worked hard to prepare a high-strung horse for a point-to-point race at Burghley. The horse was of fine quality, and Martin was sure that together they could win the race. Because of the nature of the event, however, with dangerous obstacles and some difficult jumps, William Exeter asked his elder son to ride the horse—which David did, coming home the winner.

Martin did not spend all that much time with his older sister Winifred when they were young, partly because she was six years his senior, and also because he was of a quieter nature—Winifred, like David, being energetic and extroverted. A close bond did develop, however, between Martin and his younger sister, Romayne. He used to call her "Baby," and as a child shepherded her around and cared for her with great zeal.

By the age of six, Martin was being taught piano and other subjects by a governess, Miss Winterbotham, otherwise known as "Miss Win." (Some years later, he would receive a few lessons on the piano from the celebrated conductor, Malcolm Sargent, then a young, impecunious musician living in Stamford). Following the outbreak of war, Lady Exeter had converted the Orange Court at Burghley—such a fine place for riding tricycles and pedal cars—into a hospital. With characteristic drive and sense of duty, she began organizing fund-raising events in aid of the war effort, sometimes calling upon the musical talent which she had detected in her younger son. As the guns roared and men died in Flanders fields, Martin, wearing a miniature khaki uniform, sang a new song in the Great Hall at Burghley entitled, "I Want to be a Soldier."

On Sundays the Exeters attended service at the family church, St. Martin's, in Stamford, where many forebears are buried. In the afternoon they observed a different kind of tradition, visiting points of interest in Burghley park. Setting out from the south entrance of the house, they would board a small ferry and cross a lake constructed by the famous landscape artist, Capability Brown, in the eighteenth century. Their first stop was the dairy farm, where Mr. Rowebottom was in charge. He had the habit of milking cows from a standing position, head bent down and resting against the cow's side. Over the years his posture had gradually changed, so that he always looked as if he was ready to lean against a cow and milk.

From the dairy farm, the Exeters moved on to the head keeper's cottage, where they received a warm welcome from Mr. Pepper and his dogs. Robert Pepper had inherited his position from his father, and would pass it on to his son. With his assistants, he looked after all the game on the property, such as pheasants, partridges, and deer. When Martin was old enough to learn to shoot, Mr. Pepper took him out and taught him to use a shotgun properly. From the head keeper's cottage the party walked through the bracken to the top garden and then back home around the lake. The top garden was a highlight of the walk for Martin, with its profusion of apples, cherries, grapes, gooseberries, and other fruits. Once he and a young friend made an unofficial visit to the gooseberry beds, crawling under the netting that protected them from birds—if not small boys. They emerged in due course, feeling full and very pleased with themselves. But as sometimes happens in such instances, there was a price to pay. By the time he sat down to dinner, Martin was feeling far from well. But what was he to do? Dinner at Burghley House was a formal, almost intimidating occasion, and the thought of excusing himself from the table scarcely entered his head. He simply sat there, feeling more and more queasy, until finally he was sick all over his plate. The matter was taken in stride, of course, in proper British tradition. Westley and his uniformed footmen moved in, and conversation continued as if nothing had happened.

Another time Martin put on Oliver Cromwell's boots while playing with his cousin, Gervase Falkiner. Cromwell had shot some cannonballs at Burghley during the English Civil War, and later—according to the story—came to dinner there, leaving his boots behind. Falkiner had got hold of a sword, and while the two boys were playing he accidentally made a cut in one of the boots. It was a delicate situation considering their historical value. The boys rubbed some dirt into the cut to make it look as if it had been there a long time. "A week or two later," Falkiner recalls, "Aunt Myra was showing the boots to someone and I heard her say, 'Look, you can see a sword mark.'"

Martin was nine years old when the Armistice was signed in November 1918, and was attending Lockers Park, a private preparatory school that groomed future leaders of England. Later most of the boys would go on to Eton, Harrow, or one of the other famous public schools. A few, like Martin himself, would attend the Royal Naval College at Dartmouth. Earl Mountbatten of Burma had followed this route a few years earlier.

It was no small adjustment to leave the warmth and protection of Burghley House and become an insignificant person in a boarding school like Lockers Park. T. W. Holme, the headmaster, noted in his first half-term report to Lord Exeter, "It was difficult at first to get smiles from Martin, but latterly he was much more cheerful." The teachers were decent and kindly on the whole, but did not stand for much nonsense. One, Mr. Fisher, suffered from arthritis and had difficulty holding a piece of chalk (though no difficulty in throwing it). He liked to stand behind a boy while he was working and watch his progress. If he thought it necessary, he would clip the lad over the ear with a gnarled hand, which seemed to work wonders in increasing his application.

As a member of the school choir, Martin sang lustily at concerts and other events. He was an average student, but his care and single-mindedness earned him a respectable level of academic achievement, including the Samuelson Prize for French, and also a place on the school soccer team, with all the prestige that that entailed. In his last term he was appointed Head Boy. Mr. Holmes summed things up in his last report: "Martin has worked well and has risen very creditably to his lofty position as Head of the school with its varied responsibilities."

On December 21, 1922, while Martin was out shooting in Burghley park, word reached Burghley House that he had passed his entry examinations for Dartmouth. His horizon was about to expand significantly.

5.

THE ABILITY TO OBEY

As the second son of an aristocratic British family, Lord Martin Cecil, as he was formally known, had three choices open to him when it came to choosing a career—the army, the navy, and the Church. The Church didn't appeal to him, he didn't think much of the army either, so it was all very simple. Besides, he liked the look of a naval officer's uniform.

The Royal Navy took him to interesting and colorful places like the French Riviera, Greece, and Egypt, and taught him practical skills, such as some engineering and mechanical drawing. But the most important thing he learned at Dartmouth and at sea was more discipline. The navy provided him with a higher education in one of life's most important arts, the ability to do what needs to be done—to obey.

Such training helped him to handle the challenge of a pioneering life in the Canadian West, and in due season also helped him to heed and follow the internal compulsion of spirit. As he would remark many times in later life, it is impossible to obey that "inner voice" if one has not first learned to obey in an external sense. The spirit incorporates in its makeup elements of control and design that are most specific. They cut across customary human attitudes and opinions. It takes more than a sincere love for other people, for example, or for a spiritual ideal, to express spirit accurately. It takes a willingness to change one's mind or one's behavior, perhaps even to be corrected by another person. It takes the ability to listen, to be still inside. It takes the ability to obey. All these things Martin learned in good measure in the Royal Navy, together with another useful quality, orderliness. When he went to bed at Dartmouth his clothes had to be folded just so—and positioned just so. There was a proper place for his shirt on the chest at the end of his bed; a proper place for his socks; a proper place for everything. Such an approach made for speedy dressing in the morning, which was important. It also taught him to be accurate, and that, when he went to sea, was even more important.

Self-conscious but proud in his new cadet's uniform, Martin, now aged thirteen, posed for pictures at a photographer's shop in Stamford. The next day, January 16, 1923, he and his mother journeyed to London. It was the eve of his departure to Dartmouth, and after an early dinner the two went to see *Tons of Money* at the Aldwych Theatre, where he "laughed until he cried," Lady Exeter noted. Next morning, in one of the thick yellow fogs that were then a common occurrence in London, they went to Paddington station, where many other cadets were gathering to catch the train to Dartmouth. These included Michael Culme-Seymour, whose home was Rockingham Castle, not far from Burghley. The two young men already knew each other and were to become firm and lasting friends.

Dartmouth is a picturesque little town near the mouth of the River Dart in Devonshire. Richard the Lionheart set sail from here for the Crusades in 1190. In 1620, the *Mayflower* and *Speedwell* lay off one of the coves for a week before sailing for America. The Royal Naval College, opened in 1905, nestles in the hills surrounding the almost landlocked estuary. After Lockers Park, it seemed enormous. Martin joined forty-seven other cadets in Exmouth term, as his class was called. He found the pace demanding, mentally and also physically. Every time a new cadet passed one of the dozen or so gunrooms—headquarters of the various terms—discipline required that they do so on the double.

Cadets stayed in the same term, or grouping, throughout their time at the College. As Sir Michael Culme-Seymour explains, "The fifty of us who were in Exmouth term remained absolutely separate from any other term, even the boys of the term above us or the term below. You would never meet them unless they happened to be a brother or a cousin or something like that. All your lessons as well as your games and your eating and sleeping were done with the same fifty boys. So it was natural for Martin and me, because we knew each other to start with, to become friends."

Culme-Seymour, who in later years became godfather to Martin's son, Michael, remembers Martin as being practical and very stubborn. "It was no good trying to persuade him to do something if he didn't want to do it," he recalls. "And you couldn't dissuade him from something that he did want to do. We were both quiet by nature and enjoyed going for long walks together—often in complete silence. But while I was very good at games, he had a bent in other directions. He was an excellent runner, and good at engineering and making things. There was a practical aspect running through everything he did—quiet, effective, wanting to do things and get them right." While Martin did not realize it at the time, his walks with Culme-Seymour played an important part in his future development, giving him opportunity to commune with spirit and sense an inner peace and tranquility.

Discipline at Dartmouth was clear-cut. Cadets received a "tick" for minor offences. After three ticks, they received a beating from the Cadet Captain (there were two of these to each dormitory). Typically, this would be three to six strokes with a birch rod, handed out in the washroom area when cadets were clad in pyjamas and so had little protection from the unkind hand of fate. The ultimate punishment was a public beating in the gymnasium. In his last term Martin became a Cadet Captain himself, responsible for disciplining others.

A sparse social life enlivened the regimen of the College. Dances were held in a hall called the Quarterdeck—but while parents and sisters sometimes appeared, often cadets had no one to dance with but each other. Holidays were always welcome. At the end of his first year Martin returned home for Christmas and plunged happily into the traditional activities of shooting and hunting. As usual, a Christmas tree was erected in the Orange Court and a dance was held in the Great Hall for the staff and their families.

Most young people go through a process of disillusionment at some point in their lives as they look around and see the obvious hypocrisies and contradictions of human existence. All too often, the best that well-meaning friends or relatives can do is to try to persuade them to accept the illusions. Jesus, we are told, sensed at the age of twelve the shallowness of the usual human goals, and the fact of a transcendent purpose in life. "Wist ye not that I must be about my Father's business?" he enquired, to the consternation of his elders.

A traumatic moment in Martin's own awakening came at the age of fifteen when he was confirmed in the Church of England. He prepared for the event with keen anticipation—something exciting was going to happen here that would enlarge and even transform his life. No such thing occurred. As the cycle of preparation proceeded, he began to see that the chaplain who was assisting him, while he was a nice enough person, knew little more than he did about the deeper questions of human existence. More than that, Martin realized, everything was being done by rote. The passages which he was required to memorize all came out of the past. There was no opportunity for anything new to happen. The confirmation service itself was a bitter disappointment that left him permanently disillusioned with orthodox religion.

Martin and Michael Culme-Seymour graduated from Dartmouth in 1926. With three other midshipmen from Exmouth term they sailed on the battleship HMS *Malaya* to Malta, where they transferred to HMS *Warspite,* flagship of the British Mediterranean Fleet. Martin was seventeen-and-a-half, and would spend the next three years in the Mediterranean.

The Royal Navy at this time was still the most powerful in the world, but its strength was ebbing steadily. England had assumed that she would not be involved in another major war for at least another ten years, and disarmament was an increasingly popular concept.

Midshipmen were a lowly form of life in the navy. They

were called "snotties" (the word means "offensive," or "impudent") to remind them of this fact. One of their duties was to keep a log, so as to help train their powers of observation. The first entries in Martin's journal read:

Monday October 25: Joined Warspite at 13.30 from Malaya, with the other Warspite cadets, having spent four weeks taking passage in Malaya, a fortnight of them being spent at Gibraltar. We were given the rest of the day in which to sling our hammocks.

Tuesday October 26: Slipped anchor during the forenoon and proceeded to sea, where we did 6" full calibre target shooting with Resolution and Royal Oak. I was stationed at P1 to watch the gun in action. Dropped anchor outside the breakwater at noon. In the afternoon the four of us were shown over the ship. Leave from 15.30 to 1900.

Typical of the Fleet's peacetime routine, HMS *Warspite* left Malta early in 1927 on a cruise to Mediterranean ports. Such cruises combined rigorous maneuvers and exercises at sea with equally demanding socializing ashore. Arriving at Toulon, *Warspite* fired a twenty-one-gun salute before receiving a delegation of French naval officers. That evening, the French flagship *Bretagne* held an "at home." Next day, the Commander-in-Chief of the Mediterranean Fleet, Admiral Sir Roger Keyes, gave a luncheon party aboard *Warspite;* the flagship held an "at home" in the afternoon, and the day finished with a dinner and dance ashore, and a gala night at the opera.

En route to Naples, *Warspite* lost a man overboard. After prayers, Martin noted in his journal, the captain made a speech on the quarterdeck. "... he pointed out that if ever one fell overboard one should not be frightened of the screws, which tend to throw one up rather than suck one down . . . " Martin found Naples "dirty and disappointing," but enjoyed a visit to Capri with Lord Louis Mountbatten on Mountbatten's yacht, *Shrimp*. Mountbatten served with the Mediterranean Fleet for several years during his meteoric rise to become Chief of Combined Operations and one of the three Supreme Allied Commanders of World War II.

In due course Martin transferred to the battleship *Queen Elizabeth,* which had become the new flagship of the Fleet. During a visit to the French port of Villefranche, an unprecedented event happened. Martin fell in love. He and some other midshipmen had accepted an invitation to a dance. It was, on the face of it, just another social fling ashore, but during the evening Martin met a striking young Hungarian girl who was tantalizingly different from anyone he had known before. Petite, dark-haired, with big brown eyes, Edith Csanady de Telegd, of Budapest, stirred his immediate response and interest. On April 23, 1929, elated and yet heartbroken too at having to leave the port, he wrote sorrowfully to his mother:

We have left Villefranche at last after, as far as I was concerned, a wonderful time. I am very, very sleepy after all my tremendous dancing. I fell in love last Thursday night with a sweet little Hungarian called Edith whom I met at Baroness Orczy's dance. Unfortunately I don't suppose I shall ever see her again, anyway for years and years! She is the niece of Mrs. Grant Richards (also Hungarian), the publisher's wife.

The attraction between Martin and Edith proved to be mutual. Martin wrote her a letter—she was staying with her aunt at Monte Carlo at the time—and they began to correspond.

One of the duties of a junior midshipman was to serve as

the captain's "doggie," or dogsbody, running errands and taking care of any other tasks that he might want done. While the *Queen Elizabeth* was berthed at Malta the captain called Martin into his office one morning and gave him a job of great delicacy. His new car had arrived from England. Would Martin collect it from the Customs office and drive it to the captain's house some five or six miles away on the other side of the harbor?

It was delicate because Martin had never driven a car in his life. However, to have said, "No, sir," or, "But I don't know how to drive," would have been unthinkable. He saluted and took his leave, doing his best to ignore the feelings of panic that were partying in his solar plexus.

Getting the release from the customs officer was easy, but then there was the problem of getting this shiny, immaculate motor car out of the customhouse shed, which was a stone building with a narrow exit. He climbed in and studied the controls. He turned on the ignition. Finally, very cautiously, he inched his way out of the building. He had, of course, ridden in cars before and had some idea of how they operated. Emerging unscathed from the customhouse, he navigated a succession of narrow roads and lanes filled with pedestrians, bicycles, animals, carts, and other impediments. Determined that not a scratch should mar the new vehicle, he crawled along at a few miles an hour, ignoring the honks of other motorists and the stares of passers-by. It took a long, long time—or so it seemed—but he delivered the car safely, returned to his ship, and reported nonchalantly to the captain that the matter had been taken care of.

Following a stint on a destroyer in the summer of 1928, Martin returned to England to be the best man at his brother's wedding in January 1929. He was now giving serious thought to his future. The newly developing arm of naval aviation appealed to him—he had flown on one or two flights while stationed at Malta—but he was hesitant about continuing in the regular navy. In the spring of 1929 he wrote to his mother to say he was applying for the Fleet Air Arm. "This will not be absolutely binding," he said, "but allowing that I remain in, it is the only thing worthwhile as far as I am concerned—the air is the coming thing. In fact it has already come, but still is only in its youth. Each year, since I have been here, the aircraft and their carriers have played an increasingly important part in the fleet exercises."

Indeed, naval aviation was in its youth. It was only a year or two since Lindbergh had flown alone across the Atlantic. Myra Exeter greeted her son's news with a distinct lack of enthusiasm. However there was another possibility. In 1912 Martin's father had purchased a cattle ranch at 100 Mile House, in British Columbia, as an investment for the future. He had visited the property a number of times, and kept in close touch with his agent, but he was keenly aware of the difficulty of operating the ranch effectively as an absentee landlord. As Martin's future in the navy came increasingly into question, the thought presented itself: how about Martin going to 100 Mile House to develop the potential which was undoubtedly there?

The decision was made. On the face of it, Martin was giving up a safe and promising career after having accomplished the hardest part of his naval training. He was also giving up everything else which was his by reason of his upbringing and heredity. As he remarked many years later, "Everything was just beautifully set for what should unfold in the days to come, if I continued on in the usual, traditional way. But this was the thing that troubled me, because while there was everything there apparently, I knew that it wasn't me. The easy way, in a sense, in such case is to accept the status quo and stay with that, but there was a compulsion in me which in effect said, 'You need to find yourself, and you can't find yourself if you're wrapped in all these swaddling clothes.'"

He spent his last few months in the Royal Navy studying at the Greenwich Royal Naval College (and drawing caricatures of the professors). Then, retiring as an acting sub-lieutenant, he set sail with his father for Canada.

6.

INTO THE UNKNOWN

Waving a last farewell to his mother and other members of his family standing on the dock below, Martin Cecil stood with his father at the rail of the *Duchess of Richmond* as the Canadian Pacific liner steamed out of Liverpool harbor on a grey, windswept morning in April 1930. At twenty, Martin was a handsome young man with clear blue eyes and strong, aquiline features. The breeding and tradition of many generations of English aristocracy showed clearly in his speech and manner. He had never been to Canada—and knew nothing of cattle-ranching—but he was taking his move into the unknown in stride. The British avoid getting excited about anything anyway, and the navy had done a good job of teaching him to take things as they come.

The voyage took about eight days. The ship rolled atrociously as she plowed through the cold Atlantic seas. The motion didn't bother Martin, old seadog that he was, but it discomfited many other passengers, especially when some furniture began sliding across the floor of a lounge one afternoon. There were several young people aboard with whom to pass the time—including an American girl from an intriguing place called Walla Walla, Washington—and Martin became a formidable opponent at checkers.

The trip gave Martin and his father a chance to get to know each other in a new way. In one intimate conversation, Lord Exeter confided that he knew he tended to shout when he felt frightened or insecure.

The two enjoyed a brief visit in Quebec city, strolling the old, cobbled streets, before continuing to Montreal, where they boarded a Canadian Pacific trans-continental express. On April 27, 1930, as they crossed the endless prairies of Manitoba and Saskatchewan, Martin celebrated his twenty-first birthday. The train worked its way through the Rocky Mountains and reached Kamloops, in south-central British Columbia, where Lord Exeter's agent, C.G. Cowan, waited to greet the two arrivals.

Originally from Londonderry, Ireland, Cowan served in the North West Mounted Police and fought in the Boer War before settling in British Columbia. He was a big game hunter, collecting trophies for private buyers in England and elsewhere. He also was active in the land business, and it was through him that Lord Exeter had bought the Bridge Creek Ranch at 100 Mile House in 1912. They maintained a close connection from that time on. Once in a while, Lord Exeter would come out to Canada to see how his ranch was doing.

Occasionally Cowan, a born storyteller, would visit the Exeters, and hold the guests at Burghley House spellbound with his tales of adventure in the Yukon.

A blue McLaughlin Buick waited to take the party to 100 Mile House. Lord Exeter sat in the front with Cowan's chauffeur, Bob Miller—whose wife was said to be one of the best cooks in the Cariboo—while Martin sat in the rumble seat with C.G. Cowan. The drive made a strong impression upon him—particularly, as he would remark later, upon his buttocks. It was a twisty, bumpy, and dusty journey, and took five-and-a-half hours. After passing through the small hamlet of Cache Creek, the car swung almost due north and began climbing as it entered that region of the B.C. interior known as the Cariboo. The Cariboo is a plateau some three to four-thousand-feet high, located between the Coast Range and the Rockies. As Bob Miller puttered through Clinton, some fifty miles south of 100 Mile House, and continued north along the Cariboo Road, Martin relished the beauty of the unspoilt country that was to be his home. Evergreens bordered the highway and reached as far as the eye could see—massive stands of fir and pine that cloaked the hills and gave way to fat, rolling meadows that made this one of the prime cattle-growing areas of the province. Here and there poplars fringed the shore of a lake, and he saw the split-pole fences built by the early pioneers (they have the advantage of not requiring post-holes, which are difficult to dig in hard, frozen ground). At times, his eye would catch sight of distant mountain ranges, the sun gleaming on their snow-covered peaks.

Cowan's car followed the same route as the miners and prospectors who hurried north to Barkerville in the 1860s to join in the Cariboo gold rush. A mile or so south of 100 Mile

House, the party passed the spot where a robber held up the BX stagecoach in 1890. Descending the long hill, Martin caught his first glimpse of Bridge Creek valley; a little later, 100 Mile House itself came into view, so named because it was one hundred miles from the start of the Cariboo gold trail at Lillooet. Prior to the Cariboo gold rush, however, 100 Mile House was known as Bridge Creek, and was a favorite watering hole for fur traders who moved into the interior of British Columbia in the footsteps of explorers like Alexander Mackenzie.

Bob Miller drew up beside the wooden stopping house built in 1862 to accommodate the miners and adventurers heading north to the goldfields. Now the headquarters of the Bridge Creek Cattle Ranch, it was a forlorn and dilapidated structure. Even in its heyday, according to one report, the old 100 Mile House was distinguished by its "crooked doorways, massive stove, a rickety staircase, and windowless bedrooms." A greater contrast to Martin's ancestral home would be hard to imagine. But it was here, in this unspoilt, pioneering country, that he was to find his destiny. Like Buddha, who left his father's palace to find peace and serenity under a pipal tree, or like a modern-day Thoreau, Martin confronted life at the level of its essences. It was fortuitous that his father had displayed a certain courage and vision when he purchased his property at 100 Mile House. Because he did so, there was a place to which Martin could come: a habitation in the wilderness.

All the qualities that had been required of Martin's father in caring for the Burghley estate were now required of Martin. As manager of the Bridge Creek Cattle Ranch, he was responsible for fifteen-thousand acres of deeded land and eight hundred head of cattle. He too would have to learn to work with people and inspire them to give their best. In some ways, he would have to do more than his father did, because in England the physical pioneering had all been done long ago, while 100 Mile House was virtually a blank; there was nothing there. Whatever happened would be entirely up to him. There were a few helpful neighbors around, like the Cowans, and he kept a close connection with his father by correspondence, but essentially he was on his own.

He rolled up his sleeves and went to work. This was not just a managerial position. There were ditches to dig, meadows to irrigate, cattle to round up and brand. It was a hard, active life, in a land of great contrasts. In summer, the heat could make the strongest man wilt. The long, cold winters could dip to 50 below zero Fahrenheit. And of course, the Great Depression was well underway.

7.

"SIMPLIFY, SIMPLIFY"

Canada had welcomed 1930 with a curious mix of apprehension and optimism. The collapse of the New York stock exchange in October 1929 ruined many and brought misfortune to thousands more. But even though construction and exports dropped dramatically, most Canadians believed the assurances of the experts that things would soon improve. After all, they enjoyed the second highest standard of living in the world; and in many ways it was a time of great progress as technology continued its triumphant advance.

By the time Martin and his father reached the Cariboo, however, the Great Depression was hurting more and more people. Forty percent of Canada's exports had been going to the United States. Now, seeking to protect its own interests, the U.S. moved to shut out Canadian goods. Hundreds and soon thousands of Canadians were fired from their jobs with no severance pay or unemployment insurance to fall back upon. Hope turned to bitterness for all too many young people who graduated in June 1930 from high schools, universities, and colleges, only to find there was no work for them in their chosen professions. Men who were heads of successful companies—even millionaires—awoke one morning to find themselves bankrupt. People who had never asked a favor in their life were forced to accept hand-me-down clothing for themselves and their families.

True, many Canadians, particularly middle-class urban families, lived through the Depression relatively unscathed; for them it was a time of adventure and change, with exciting developments in such fields as air travel and the long-distance telephone service. But prairie farmers who saw wheat prices tumble and topsoil succumb to drought and winds, despaired. So did a swollen working class whose services were no longer needed. Hitherto law-abiding citizens took to the streets, marching, shouting, and protesting. Long lines of sad-faced men queued at city soup-kitchens or huddled together against the wind as they rode the country on freight trains looking for jobs that simply did not exist. Once, a small group of men hiked eleven miles when they were told there was a job on a farm: they arrived to find that forty others had been there ahead of them. "I will end unemployment or perish in the attempt," proclaimed R.B. Bennett, elected prime minister in the fall of 1930 when it became evident Mackenzie King had no answer to the crisis engulfing the country. But it was not long before Bennett too was the target of resentment and frustration.

The Depression stripped away material wealth, possessions, and status in a more or less arbitrary way. People didn't have the luxury of making a choice. Those whom it hit took the view—understandably enough—that disaster had struck; some threw themselves out of windows. Yet it's interesting to consider that some people throughout the centuries have voluntarily given up what seemed like material safety and comfort. They had a suspicion that society as a whole was deceived; that true security and comfort do not lie at the material level at all. "Our life is frittered away by detail," wrote Henry David Thoreau, the man who lived by Walden pond. "Simplify, simplify." And again, "Most of the luxuries, and many of the so-called comforts, of life are not only not indispensable, but positive hindrances to the elevation of mankind."

Whether it is necessary to give up material things in order to find "enlightenment" is a debatable question. It would seem that what most needs relinquishing is the more intangible clutter that accumulates in human minds and hearts and obscures the light of spirit—the clutter of prejudice and resentment. However, there is no doubt that Thoreau would have approved of the spartan quarters which Martin inhabited for his first two years in the Cariboo.

He lived in a tiny room in the old 100 Mile House stopping place measuring eight feet by nine feet. There was no butler here to advise him that dinner was served, no maid to make his bed, and no one to polish his shoes; but then there were not many days when he needed polished shoes. As for the diminutive size of the room, this was an advantage, at least in winter. It made the room easier to heat. He had a cot to sleep on, and for heating, a small cook stove with a firebox about thirty inches by twenty-four inches, of a kind used by shepherds when they herded sheep in the summer. The stove delivered heat quickly, but it also went out quickly. The room would heat up to about a hundred degrees, but an hour or so later, on a cold winter's night, the temperature would be back below zero. Martin developed a special technique to combat this situation. Before retiring, he would make sure that he had plenty of newspaper and kindling within reach. In the morning—when he had summoned enough courage—he would extend an arm swiftly and efficiently from under the covers, insert the kindling into the firebox, and having applied a match, luxuriate under the covers until the room was bearable.

The central portion of the old stopping house was a story and a half high. During the Cariboo gold rush of the last century, thirsty miners on their way to the creeks of Barkerville crowded into the bar which occupied a large part of the ground floor. A loft above held sleeping accommodation. Later, this area was divided into cubicles and two two-story wings were added, one on either side of the building. The stopping house did a thriving business in its day, and may have been a richer goldmine than the diggings further north. The great pack trains and freight wagons all stopped there on their way to the goldfields, either for a meal or for the night. So did the smartly painted Concord stagecoaches, with their yellow and black running gear and white canvas tops.

As the gold rush subsided, ranchers began to settle the surrounding area, which kept trade coming into the place. Instead of miners and prospectors, cowboys appeared, and stock saddles took the place of pack saddles on the ponies standing wearily at the hitching rails in front of the old building. The stagecoaches made their final trip over the historic route in 1917 and 100 Mile House saw them no more. They were replaced by motor cars, and dust-coated, begoggled chauffeurs took the place of the stage drivers.

In addition to the stopping house, the Bridge Creek Ranch also included a general store with post office, a large barn, and a fine log building that was used as a carpentry and blacksmithing shop. This was built around the turn of the century by Syd Stephenson who, with his brother Frank, owned the property at 100 Mile House prior to its purchase by Lord Exeter. (The Stephensons were reportedly descendants of George Stephenson, the inventor of the steam locomotive; they gave the ranch its existing brand, a quarter circle over a reversed "S," suggesting a bridge with a creek running beneath.) With a population of ten to twelve people, including ranchhands, the storekeeper and his wife, and also a government telegraphist and his housekeeper, this was 100 Mile House in May 1930.

William Exeter stayed at 100 Mile House for a month, looking things over with Martin and helping settle priorities. They decided that the first necessity was to build a modern hotel or lodge that would attract the traveling public and put 100 Mile House on the map ('modern' meaning that there would be running water and at least one bathroom). For certainly no one in his right mind would think of staying in the existing rundown structure.

While Martin had no experience whatsoever at building, he had brought a book with him from England that gave some encouragement for the challenge ahead. Entitled *Every Man His Own Builder,* it had been published in London in 1912. Said the author in his preface:

In the following pages I have endeavoured to show how any man of normal bodily strength can at need build his own house without the aid of skilled labour. To the ordinary Englishman the idea of building houses without skilled artisans no doubt appears strange, but to our colonists, who in out of the way places and for various reasons—often financial ones—are frequently compelled to do so, it will appear far less strange. It is my hope that when he has read this book the idea of the impossibility or incongruity of building a house with other than skilled labour will also have disappeared from the most stay at home and old-fashioned Englishman's mind.

With no one around to help him, Martin depended upon such books for instruction. He enrolled in a correspondence course on architecture and bought manuals on plumbing, wiring, and other subjects, studying at night by the light of an oil lamp as the work progressed. The first step, drawing some plans, he found easy, as he had learned mechanical drawing in the navy and had a natural feel for it. He discussed his plans with his father, and together they chose a spot for the new building and laid it out on the ground.

At the end of May, Lord Exeter returned to England, while his son continued the work of building the new hostelry and running the ranch. The lumber for the lodge came from a small sawmill which Syd Stephenson had erected a mile or two away on Bridge Creek; it was powered by a water-driven turbine located beside a waterfall. One of the ranchhands, a West Indian, was a resourceful man and had the job of operating the mill, though Martin ran it himself on occasion when he had mastered its idiosyncrasies. The sawmill needed a good volume of water in the creek to run properly, and unfortunately these were dry years. The result was that the boards tended to be uneven because the power would start to give out about halfway through the cut. Not only would the lumber that emerged be of different sizes, but each board would also be thicker at one end than the other.

There was no planer at 100 Mile House in those days either, so the lumber was indeed rough—just like the carpenters. Nevertheless the building went up and stayed up, despite some tense moments, and pessimistic comments from observers. Once, the walls began to bulge under the weight of the roof that Martin was putting on. His books had not prepared him for such a catastrophe, but he did some quick thinking and hit upon a solution. He collected some haying gear and rigged up a block and tackle arrangement between the two walls, with one end of the steel cable attached to his Model A coupe. He then pulled the walls together and spiked a couple of jackpines across the top as joists. They are still part of the structure to this day. It took two years to complete the 100 Mile House Lodge, as it was called; because of the severe winters, the work could only be done in the summer.

As it turned out, the heating in the Lodge was not much better than in the old 100 Mile House—perhaps because there was no insulation available other than wood shavings. But there was running water, and there was also electricity. Compared to other accommodations on the Cariboo Road the 100 Mile Lodge was a luxurious and ultra-modern place to stay.

Devising a water system worthy of the new building had also posed a challenge. When Martin first came to 100 Mile House, anyone needing water simply scooped it out of Bridge Creek in a bucket, put it in a drum, and hauled it home. There was a well, but nobody could drink from it. The water was no use for washing either because it was so hard. Speaking to the 100 Mile House Rotary Club in 1984, Martin described how sixty-eight years after one Thomas Miller built the original roadhouse, running water came to the little community.

The village as it is today sits right on top of what was the closest hayfield we had on the ranch. It was an irrigated hayfield. The water came down from Bridge Creek; it was taken out of Bridge Creek close to the Horse Lake bridge and came through ditches and flumes all the way down to this field. We always had a good crop of hay here, which we lost when the village was built on it. But at that time it didn't look much like it does now, because there has been a lot of levelling done as the village grew.

There were gullies; there was one quite deep gully where Herb Auld's garage now is, and that was a perfect site to bring water out into a flume—this was irrigation water—and into a hydraulic ram. This was a little ram that tick, tick, ticked away day and night. At least, it was supposed to, until it got plugged up occasionally by frogs or fish or something else. And it would pump water across the Cariboo Road up into a thousand-gallon wooden tank that I put above the Lodge. That was the water supply for the Lodge. Of course a thousand gallons doesn't go very far and a little hydraulic ram doesn't deliver very much water at a time. The water would go down during the day and then the ram would pump it up at night, that is if it continued to work. Well naturally one became very much attuned to that little far-distant tick, tick, tick, and if it stopped, I would have to go rushing out and get it going again.

The system only worked during the summer, of course. It had to be drained before the winter freeze, at which point water would again be hauled from the creek. Martin put plumbing in the Lodge and installed a 32-volt electrical system. This was strictly for lighting. As he told the Rotarians, there was "no such nonsense as washing machines or refrigerators or any of these things." There was ice, however: this he cut out of Exeter Lake during the winter and stored in icehouses at the Lodge. "It lasted pretty well all summer," said Martin, concluding his talk. "Because we never had anything else, because until then it was all oil lamps or gasoline lamps and hauling water and all the rest, this seemed like heaven. Wonderful! The simple pleasures of life!"

8.

COWBOY

For Billy Roberts, a grizzled, old-time cowboy, it was all highly humorous: but for Lord Martin Cecil, a young greenhorn just out from England, his first horseback ride in May 1930 was highly unamusing—except in retrospect. Shortly before his father returned to England, Martin and Billy Roberts went looking for cattle on some nearby range. It was a beautiful day when they started out, and Martin wasn't too concerned that he didn't have any rain gear. Before long, however, the sky changed dramatically and unleashed large quantities of snow. Every time Martin rode under a branch a further dollop fell down his neck. Finally, as they trotted home along the Forest Grove Road, the snow turned into drizzle, which finished the job of saturating him and his old tweed jacket. Billy Roberts, properly outfitted with slicker, chaps, and everything else a cowboy is supposed to wear, was chuckling to himself the whole way home ...

Another early misadventure related to sheep, which were a big problem in the Cariboo during the Depression. There was never enough feed for them, either in summer or winter, and at a time when four cents a pound was a good price for cattle, their value was marginal at best. During the same spring of 1930 Martin joined in a search for sheep range at Windy Mountain, some forty-five miles east of 100 Mile House. It is not so much a mountain as a large rock projecting out of the ground. An Indian guide named "English" Decker had spread the word that there was some good sheep pasture there, and an expedition was organized with English Decker as guide; Alvin Miller, a young Englishman who was the bookkeeper at the Highland Ranch; Alan Mackenzie, a shepherd; and Martin.

There were no airplanes in the Cariboo in those days. It wasn't possible to go up and fly around for half-an-hour to see what the country looked like. Impressed by English Decker's glowing account of the fine sheep pasture they would find, the expedition set out one morning at the beginning of June with high expectations, their pack animals loaded with provisions and camping gear for the four-day round trip. Again, the weather turned wet, with both rain and snow at times.

On the second day, Martin could see Windy "Mountain" through his binoculars. "Where's the sheep range?" he asked, since all he could see was the rock with a lot of scrub surrounding it.

"Not far now," English Decker replied. Occasionally someone else would ask the same question, and the reply would be the same: "Not far now."

They reached the foot of Windy Mountain.

"Where's the sheep range?" everyone asked.

"Oh," English Decker replied, "no sheep range out here." Suppressing a desire to shoot their guide on the spot, the party deliberated briefly and finally turned round and headed for home. Presently English Decker spoke up again.

"Sheep range on other side of mountain," he said. But everyone had had more than enough by then, figuring that either he didn't know what he was talking about or he was merely trying to earn some money for himself during hard times.

Immediately to the north of the Bridge Creek Ranch lay the thirty-five-thousand-acre Highland Ranch, owned by another titled Englishman, Lord Egerton of Tatton. With extensive interests in Kenya, Lord Egerton came to the Cariboo but rarely. When he did visit, after an absence of several years, he always looked for his favorite hat—raising an uproar if it was not hanging on the hook where he had left it.

C.G. Cowan, who supervised the running of the Highland, also owned two large ranches himself, the Onward and the 150 Mile. They are located east of the town of Williams Lake, which is itself sixty miles north of 100 Mile House. Cowan and his wife Vivian extended a warm welcome to Martin upon his arrival in Canada, opening their home at the Onward to him. The Onward was a working ranch, but the dinner table was always set with white linen and silver, and many interesting people came to visit. Vivian Cowan had met A.Y. Jackson, one of Canada's "Group of Seven" painters, while taking an art course in Banff. She invited him to come to the Onward and do some painting in the Cariboo, which he did, with the result that the ranch became a mecca for artists—a mecca into which Martin, with his love of drawing, blended easily. Mrs. Cowan has happy memories of those times at the Onward. "Martin was quiet but he had a very hearty laugh," she recalls. "There was always a lot of laughter when we sat down together at the table. We were very fond of him. He was affectionate and kind, and used to draw pictures of animals and cowboys for our two daughters." The Cowans' older daughter, Sonja, now Mrs. Hugh Cornwall, thought the young English lord was "the greatest thing there ever was—very good-looking and full of fun." At the time, she attended a private school on Vancouver Island. Martin once drove her down to Vancouver from the Onward in thirteen hours, remarkable time considering the state of the road and the flat tire they encountered along the way.

While there were times when he felt lonely and homesick, basically Martin fitted into his new environment well. Perhaps his quiet nature helped. Rather than pushing himself forward and asking a lot of questions, he preferred to learn by watching others. Cowan wrote to Lord Exeter on July 10, 1930: ". . . it does not take long to find out that Martin is OK for us out here, and I'm sure both the Onward and Bridge Creek can never kick because he left England to make a home out amongst us. We are right glad to have him and I am so very pleased to find everyone as well as myself looking upon him as quite settled and one of ourselves. Martin comes to us in a day or two for a little rest. He is working hard, so hard that when I last saw him he not merely had his coat off, but his shirt, and the perspiration was running down him. Nice to be young!"

Angus McLachlan, foreman of the Bridge Creek at this

time, was a Scotsman who did not say much but was good at his job. He did a lot of the riding himself, and with a fifteen-thousand-acre spread, there was plenty to do. He often rode at night, keeping an eye on the cattle, which in the Depression days were a tempting commodity for someone's table. In May 1931, however, McLachlan handed in his notice, having bought the 141 Mile Ranch. No doubt the canny Scot had been saving his money for years. Martin was in the middle of building the Lodge, and was not at all sure if he was capable of running the ranch on his own as yet; fortunately he was able to find one or two capable men who knew their work, including Don Laidlaw, a bow-legged cowboy who later became his cattle foreman. Laidlaw, who only looked at home when he was on a horse, worked with Martin for many years and the two became firm friends. When someone came to Martin once and complained about Laidlaw, Martin listened quietly for a few moments and then said simply, "You've told me, now go and tell Don."

Naturally the manager of a large cattle ranch was always fair game for criticism. Once Martin was invited to a public meeting which had been arranged at Forest Grove, fifteen miles from 100 Mile House, by a group of small ranchers and homesteaders. When he arrived—just a young fellow in his early twenties—it was obvious he was the target of strong feelings of hostility and suspicion. Various ones stood up and complained about the Bridge Creek and Highland ranches, and how they were trying to squeeze the small operators.

"I've heard Lord Martin goes to Victoria and works things out to suit his own interests," said one man truculently. Victoria is the capital city of British Columbia. As it happened, Martin had not even been there at that point.

Some government officials from Victoria were present at

the meeting. "We can always tell when someone is trying to get something," one of them said, "by the size of his file. Lord Martin's file is very small indeed." So Martin had a little support from that direction.

But then a homesteader named George Hendricks stood up and in spite of the hostile attitudes of virtually everyone else in the room said, "I've been living in the middle of the Highland Ranch for many years and I've never had any trouble. Everything that needed to be discussed has always been discussed and worked out and it was all quite all right." Martin's heart went out to the man. He was no longer alone. As time went by he became more accepted in the area, but understandably he and Hendricks became good friends; Martin often used to drop in and visit him at his homestead.

During the early 1930s Charles Cowan, his health failing, asked Martin if he would take over the management of the Highland Ranch. Covering fifty-thousand acres between them, the Highland and Bridge Creek ranches were a substantial responsibility for anyone, let alone a young man who was still learning the business. He took the job on, however—not without considerable trepidation—and ran the two operations as one. Altogether there were about two-thousand head of cattle and a similar number of sheep, though as soon as possible he sold the sheep because they weren't a practical proposition. Alex Morrison, a shepherd from the Isle of Skye who had come to the Cariboo in 1929, had looked after the sheep for some years. After they were sold—being flexible—Morrison became a cattleman. He was foreman of the Highland for many years until it was sold, and later became foreman of the Bridge Creek Ranch.

The headquarters of the Highland was at the 105 Mile House, where a Chinaman known as Ah Joe cooked for the ranch crew and for occasional guests. He also had many other tasks that kept him busy, such as milking the cows, looking after a small flock of sheep, keeping a garden, doing the laundry and, last but not least, making butter. Ah Joe was a good butter maker, but it was his firm and unshakable rule to serve the oldest butter first. Since there was always more butter on hand than was being consumed, rancid butter was a way of life at the 105.

Ah Joe ruled the roost at the 105, even though the cattle foreman, who was responsible for the ranch, lived there; to prove the point, Ah Joe once chased the foreman around the dining table with a meat cleaver. Eventually, having decided that he should go back to China, Ah Joe resigned his post and disappeared from the scene. Some years later a small figure came walking up the Cariboo Road. Ah Joe had returned from China and wanted his old job back. He was hired again, and worked at the 105 for several more years until the lumber industry began to boom in the area and he left to work in a logging camp, which was the last anyone saw of him.

There was no such thing as morning or afternoon coffee breaks in the Cariboo in those days. Martin and the ranch crew would work all day and think nothing of it. He often hired men from the Canim Lake Indian reserve, twenty miles east of 100 Mile House, for haying, cowboying, and branding. They were good workers and knew the country. After a hard day at branding Martin would fetch some cold beer; although it was illegal to supply liquor to Native people, branding is thirsty work.

Martin did not keep a regular diary, but some entries he made in 1933 give a glimpse of a working rancher's life:

Thursday, July 13: Up at 0600. Sold 55 hides to Mr. Weltmay. 48 at 4 cents per lb., 7 at 1 cent (spoilt), also 20 lbs. of horse hair at

10 cents per lb. Total $42.69. Rode up to the Highlands. Howard ploughing. Hay not very good except in meadow at head of lake. Sweet clover good. Hay in meadow behind store very good at far end. Alfalfa no good. Good catch where seeded down under new ditch.

Tuesday, July 18: Took lunch and rode out to Buffalo Lake, via Willowdale, Upper Willowdale, Wallace Meadow, and Buffalo Ranch. Lower Willowdale hay very good. Buffalo Ranch patchy. Beef at Buffalo Lake doing well. Flies bad in the brush.

Monday, July 24: Worked with hay crew all day. Finished hauling behind barn and started behind store. Mowing started at lower Willowdale. 90 degrees in the shade. Hoodoo day—barn hoist rope broke three times, broke new mower knife, broke new slings, Ben fell off his stallion, Foulis had a runaway with the mail team, Roamy and Patrick, smashing harness and Democrat.

Wednesday, Sep. 22: Heavy snowfall last night and still snowing through the day. All flowers gone and our beautiful garden died in a day. The usual round in the afternoon, hotel still busy.

Polo was a favorite recreation in those early days. Martin had brought his equipment from England, and decided to see if he could introduce the game to the Cariboo. With lots of cowboys and cow ponies around, 100 Mile House, he was sure, should be an ideal place—which in some ways it was. One of the participants was a Constable Green, of the B.C. Constabulary, who used to come up from Clinton; he had served in the cavalry in India and was an excellent player. The men played on a large field west of 100 Mile House. The ground was not too even, of course, and there were little piles of extraneous matter here and there to contend with. Martin also built a golf course, having been told it would appeal to visitors. He put a lot of work into it but, as it turned out, nobody was very interested except the few sheep that remained at the ranch; they liked to bed down on the greens, which were black, composed of oiled sand.

Stampedes and rodeos were a big attraction in the area, as they still are. Reporting on the annual Forest Grove stampede, the *Williams Lake Tribune* told how Henry Bob and Clifford Eagle—both employees of the Bridge Creek Ranch from time to time—won the wild cow milking contest. Eagle subdued the cow after seizing it by the tail, which prompted the *Tribune* writer to suggest that dairymen with stubborn cows follow the same approach.

Martin dabbled at cross-country skiing during his first winter, sawing a birch board in half to make some skis. He didn't get a good bend on the front parts, so they were inclined to dig into the snow. He fastened thongs over the tops, into which he put his boots. During the same winter some of the ranchhands asked him to give them boxing lessons, thinking that because he had been in the navy he must be good at it. This wasn't the case at all, but they had an enjoyable evening sparring around together.

Life in the Cariboo was simple and down-to-earth, and difficulties were sorted out in a straightforward manner. When Charles Cowan's foreman got in a fight with one of his shepherds, the sound of their blows carried a hundred yards away. In another incident an Indian ran amok in Williams Lake and started fighting passers-by. The local policeman, Sergeant Gallagher, ran to the scene and engaged the Indian in a fist fight. The men fought three rounds. Though the sergeant knocked his man down twice, the Indian kept coming back for more, until the sergeant knocked him out in the third round with a nice right to the jaw followed by a left. Law and order having been upheld, the Indian was summoned before the magistrate and sentenced to two months in prison.

Wishing to share his new world with people in the "old country," Martin wrote an article for the prestigious British magazine, *The Field*. It is a practical, down-to-earth account of a rancher's life, but his deep love for the Cariboo—and the poetic side of his nature—shone through his words.

The day's work starts at seven, though the horses will have been rounded up from the pasture, brought into the barn and attended to before breakfast, which is at six. From the moment that the spring break-up takes place to the time of the freeze-up in the fall there is no time to be wasted. The "open" seasons are all too short! The spring is so brief that one might almost say that winter jumps straight into summer, and yet there are fields to plough, crops to put in, irrigation ditches and flumes to fix, gardens to plant, calving cows to watch, and a hundred-and-one other jobs all shouting for attention. Then comes the round-up for branding of the calves. This may take anything from two days to two weeks, according to the number of cattle and the distance they have drifted on the range.

It means long hours and tiring days in the saddle. One method of branding a calf is for a man to rope the calf by the hind legs, while another man throws it and holds the front end down—have you ever tried to throw a two-month-old calf? It isn't as easy as it looks. The calf would just as soon be on top as you would and it can kick harder than a rugby international. There are also fences to mend, cattle to pull out of mud holes, horses to break—in fact one d---n thing after another!

Midsummer is now here with its hot, cloudless days and cool, clear nights. The scent of clover and alfalfa in flower is wafted from the hay meadows on the gentle breeze and the poplar leaves are set shimmering in the sunlight. Soon the familiar sound of mowing machines is heard and the haying season has commenced. The first of the beef is now fat and ready for shipment.

When the haying is completed it is just about time to commence harvesting, and by the time the crop is cut and threshed the snow is flying and it is time for the fall round-up. There is no rest for the wicked! But there are plenty of good trout in the lakes and creeks asking to be caught if there is a moment to spare; not to mention duck, geese, different kinds of grouse and, in some parts, even pheasants for the shotgun; or moose, caribou, mule deer, bear, mountain sheep, and goat for the rifle. Then there are frequent dances and an occasional stampede or rodeo in the district.

The clear days and cold nights and the superb colouring of the countryside make the fall the most beautiful season of the year. But, like the spring, it does not last long. The winter is long and cold with plenty of snow, but an abundance of sunshine. The cattle, though they remain out in the open, have to be fed daily, each beast on an average consuming one ton of hay every winter. And so the year draws to a close. Night succeeds day; the stars look down from a cloudless sky on a silent white world. A silence as of suspense, waiting maybe for life to come to everything and another year's work to the rancher.

9.

LIFE UNFOLDING

Back of all the happenings on the planet—and despite the disinterest and generally uncooperative attitudes of much of humanity—the universal spirit continues to unfold its own coherent design and purpose: its own "implicate order," as physicist David Bohm has termed it. This invisible movement of life was back of even such a simple thing as the opening of the 100 Mile Lodge in the spring of 1932.

Martin's parents were the first guests to sign the register, after which Myra Exeter took a picture of everyone in front of the fine new hotel. Martin stood relaxed and smiling on one side of the group, wearing an old tweed jacket and open-necked shirt, hands thrust deep in the pockets of baggy work pants. Probably the whole population of 100 Mile House was present—the ranch hands, with their rough-hewn, weathered faces; the kitchen help and waitresses; Jack Lloyd, the storekeeper, and his wife; the government telegraphist; and one or two children.

Lord and Lady Exeter stayed from the end of April until mid-July. It was the year that Charles Lindbergh Jr., the nineteen-month-old son of Charles and Anne Morrow Lindbergh, was kidnapped. Amelia Earhart became the first woman to fly solo across the Atlantic. In India, Gandhi's campaign of non-violence continued to gather impetus, fuelled by the repressive measures of the British. And in the city of Nashville, Tennessee, totally unremarked by the world at large, a young man named Lloyd Arthur Meeker came to a profound recognition of the responsibility of humankind, and of himself in particular, to be the means for the action of spirit on earth. Raised in a poor cabin in the foothills of the Rockies, Meeker's background could not have been more different from that of Martin, but both men were moving with the rhythms and compulsions of that universal spirit.

By summer, people were arriving at the Lodge from all parts of Canada, the United States, and from farther afield—including a Mr. Campbell, from Shanghai; Victor Cazalet, from the British House of Commons; and Lord Brassey and his son Peter, also from England (Peter Brassey later married Lady Romayne Cecil). Later, Sir Michael Culme-Seymour would visit while serving in Ottawa for a year as aide-de-camp to the Governor-General, Lord Bessborough.

Marie Lloyd was in charge of the Lodge. She was the wife of Jack Lloyd, whom Martin had hired in 1931 to run the store, replacing an older man who had been born in 100 Mile House. It was the first time Martin had had to fire anyone and he found it a painful thing to have to do. But the job was a strenuous one, and the store urgently needed new ideas and energy.

The Lodge contained nine rooms upstairs. Downstairs there was a cozy living room, complete with large fireplace, and two dining rooms, one for guests, the other for the ranch crew. Martin had built an apartment for himself adjoining the living room. This also had a fireplace: more than once, on a cold spring night, he warmed a newborn lamb by the fire to save it from freezing to death.

The Cariboo was a trusting, open place in those days. The Lodge was rarely locked. When guests arrived they found a blackboard listing the various rooms and the prices, and if no one was around they simply chose a room, marked it off on the blackboard, and made themselves at home. Sometimes the Lodge would be empty when the staff went to bed—and full in the morning. Early risers left payment on their pillow, or at the desk; seldom did anyone leave without paying.

Martin had a standing policy during those Depression years. He would always feed anyone who needed a meal if they were willing to do some work in exchange. Since cooking and heating were done by wood, there was usually lots of work at the woodpile. Sometimes men would stay a week or longer, working for their board.

The Lodge quickly established an excellent reputation for itself (Ah Joe's bread playing its part). "It was very plush indeed, at least for those days," says Mrs. Hugh Cornwall, who stayed there often during the 1930s and used to go out and pick mushrooms for the guests' meals. "The food was excellent—steak and mushrooms, fresh strawberries from Canim lake, homemade pies and so on." Not only that. At the rear of the main building stood the "smallest beer parlor in B.C.," with room for ten or twelve men in one section and half-a-dozen women in the other.

Harriet Morrison went to work as a waitress at the Lodge in June 1934. She received $15 per month plus room and board and tips, and slept in a small room off the kitchen. Comfortable and well-appointed though the Lodge was, it lacked some amenities that would be considered essential today. Since there was no central heating, Morrison sometimes woke up to find a coat of frost on her bedcovers. Moreover, the running water only ran in the summer—the water supply was shut off in the fall so as to avoid burst pipes. This meant that in winter a person who wanted to use a toilet had to brave the frigid outdoors and journey to the outhouse.

Many visitors to the Lodge had heard about the English lord who lived at 100 Mile House and were naturally curious to meet him. However, their idea of what a "lord" should look like did not usually correspond to the fact. Harriet Morrison, who still lives in 100 Mile House, recalls a typical exchange:

"Where's the lord, miss?"

"He's in the dining room."

"Where? I don't see him."

"There. He's over there."

(Shocked pause). "But he's got jeans on."

"Yes, he's working."

A new gold strike in the northern Cariboo in the 1920s, followed by the development of a quartz mining operation, all helped stimulate automobile traffic on the Cariboo Road during the early 1930s. Such travel was infinitely more comfortable than in earlier days, but was still an adventurous undertaking—particularly in winter. When two couples from Vancouver checked in at the Lodge one day in January

1933 it marked the end of an incredible ordeal during which the party spent four days and nights on the forty-seven mile stretch of road between Clinton and 100 Mile House, and almost thirty-six hours without food. Said the *Tribune* in a front page story: "Unable to travel in high-heeled shoes when their cars failed in deep snow at the 83 Mile hill, Mrs. Hatch took them off and trudged the highway in her stockinged feet. Beyond being slightly frostbitten she was apparently no worse for her trying experience. Teams of horses were later employed to get the cars through to 100 Mile House." Martin had similar, though not such extreme experiences. Once, driving back alone from Vancouver, he skidded into a snowbank at 90 Mile one evening and discovered, to his horror, that he had forgotten to bring a shovel, a cardinal rule of winter driving in the Cariboo. Unable to extricate the car, he started walking toward 100 Mile House, only to remember, after about a mile, that he had forgotten to drain the radiator. Back he went. It was five in the morning before he finally reached home, cold, tired, and very chagrined.

Martin had been writing to his Hungarian sweetheart, Edith, ever since he first met her in Villefranche. The daughter of a senior civil servant in Budapest, Edith, too, came from an aristocratic tradition. As a child, she spent the summers in a picturesque part of Hungary near the Tisza River. There, looked after by an Austrian nanny named Josephine, she and her brother Aurel spent their days swimming and sunning on the river bank and playing with their young friends. Neighboring landowners visited her parents' cottage in their horse-drawn carriages and landaus; the cottage was set in an orchard bursting with all kinds of delicious fruits. After the

William, 5th Marquess of Exeter, with Myra Exeter in the grounds of Burghley House – 1926. Below, an 1875 chromolithograph of the historic Elizabethan mansion.

William Cecil, Lord High Treasurer and chief adviser to Queen Elizabeth I, astride his mule. Cecil helped lay the foundations of the British Empire.

Martin astride his pony. He learned to throw himself off with a minimum of discomfort if the pony got going too fast.

Above, the Exeter children: From left, Romayne, the youngest, Martin, Winifred, and David. Right, his very own lawnmower!

The soccer team at Lockers Park School (Martin standing, fourth from left).

Out shooting with his elders.
Right, as a young cadet.

Dartmouth Royal Naval College, England—Michael Culme-Seymour standing fifth from left, Martin last but one. Left and below, drawings from Martin's logbook.

Yachts off Calcara Steps

HMS Queen Elizabeth, Flagship of the British Mediterranean Fleet.
Below, the old 100 Mile House was built in the early 1860s as a stopping place for miners and adventurers on their way to the Cariboo gold rush.

The sailor turned rancher weighs anchor at the 108 Lake. Left, a small church in Surrey, England, was the scene of Martin's wedding to his sweetheart Edith in 1934. "Far away from her native land," reported the London Daily Mirror in a front page story, "a beautiful Hungarian girl today married the younger son of one of England's most famous old families."

After the christening of their son Michael, Martin and Edith pose in front of the 100 Mile Lodge with Martin's sister, Romayne, and father, William Exeter.

Watching father at work. Right, three generations – William Exeter with Martin and Michael, 1950.

First World War the family moved to Budapest, where Edith enrolled in a girls' school and, at her father's insistence, studied French as an additional language at home. As she grew older, she sometimes visited relatives in England and in Monaco—where she was staying when she met Martin.

As his friendship with Edith bloomed, Martin's thoughts turned increasingly to the possibility of marriage. He broached the idea in a letter in the fall of 1933. Edith, who had been taken with the handsome, aristocratic young Englishman from the beginning, accepted his proposal, and they announced their engagement. They were married on January 18, 1934, at a small church in Virginia Water, Surrey, England. The wedding had all the ingredients of a storybook romance.

"Far away from her native land," reported the London *Daily Mirror* in a front page story, "a beautiful Hungarian girl today married the younger son of one of England's most famous old families. The bride, slight and dark, was Miss Edith Lillian Csanady de Telegd, the twenty-three-year-old daughter of Mr. and Mrs. Raoul Csanady de Telegd, of Budapest. She married Lord Martin Cecil, the second son of the Marquess of Exeter." Martin's brother, David, was the best man and his sisters were the bridesmaids.

After honeymooning in Estoril, Portugal, the couple reached 100 Mile House in the spring of 1934. Shortly after their arrival they drove to the Onward to visit the Cowans, with whom Edith found an immediate friendship. Sonja Cowan remembers Edith as "very refined and well educated —full of fun." During her summer holidays at the Lodge, the former often joined Martin and Edith on horseback rides —Edith had been trained in horsemanship in Hungary. Another favorite activity was sailing. Martin kept a small boat at the 108, and would often take Edith and others out for a sail on the lake. Harriet Morrison, who worked in the Lodge at the time, sometimes tidied their apartment or took Edith breakfast. They became good friends, and when Morrison had twins, Edith asked if she could be their godmother. "She was always very generous," Morrison recalls. "Later on, when Alex and I were living at 20 Mile, she heard over the radio that we had been burned out. She immediately packed up a box of clothes for the children and sent it to us."

On September 1, 1935, Edith gave birth to a son, William Michael Anthony, at the Royal Inland Hospital in Kamloops. Lord and Lady Exeter and Martin's younger sister, Romayne, attended the christening conducted by the Bishop of the Cariboo at 100 Mile House. Life had taken another step in the unfoldment of its purposes.

10.

A FULL PLATE

In many ways the 1930s were a time of bare survival for Martin, as for so many others. Possibly—although his worn and patched jeans should have belied the notion—some imagined that because of his heredity he had substantial resources. Apart from a modest allowance from his father, he had no resources at all. His grandmother had invested $500 for him in a bank in Kamloops, and this enabled him to buy the Model A coupe that helped pull the walls of the Lodge into shape. But for several years he did not even draw a salary from the ranch because the money simply was not there. He had to pay others, but he could not pay himself. Every cent counted. There were nights when he lay awake for hours worrying how to handle some need or other: how to feed the sheep when there was no feed; how to repair the hay tackle when there was no money.

Inevitably, the thought of going back to England flitted through his mind once or twice. The opportunity was certainly there. At one point his brother David suggested that he come home and assist in running the Burghley estate. But somehow it never seemed a viable choice; the compulsion to stay and see things through was stronger.

Cattle ranching is an unpredictable venture in the best of times. As Nina Woolliams puts it in her book, *Cattle Ranch, the Story of the Douglas Lake Cattle Company:* "Nothing about it is constant—neither the animals, the weather, the people, the machinery, the land, nor the amount of government interference. This unpredictability makes for a business that is tough and challenging both physically and economically and that attracts and holds only those who have spirits to match. It has always been so, even though its make-up has changed often." Because of the Depression, the cattle industry in the Cariboo and elsewhere passed through an even greater flux in the 1930s than usual. As cattle prices in 1933 dropped to their lowest point since 1906, cattlemen began to realize the necessity of organizing themselves if they were to survive.

In such a context, it soon became evident that there was a use for the leadership qualities which the Royal Navy had nurtured in Martin. Against his own personal inclinations he began to play an increasingly active role in the development of the cattle industry in the Cariboo, and farther afield as well. He found this intensely uncomfortable at times—he was shy and ill at ease in public and had difficulty expressing himself before an audience. Yet that very discomfort was a catalyst in his life. As his responsibilities in the cattle industry increased, he became more and more aware of the necessity of rising above his human quirks and limitations—many of them ingrained from childhood.

The Cariboo Livestock and Fair Association had established itself at Williams Lake in 1928, quickly making a splash with its annual Fall Fair. In October 1930, local livestock producers formed a Cooperative Livestock Selling Agency, which set up a public stockyard in Williams Lake and hired an agent.

The president of the Cariboo Livestock and Fair Association in 1930 was Charles Moon, and the vice president was R.C. Cotton, who was also president of the Cariboo Stockmen's Association. Martin became a friend of these and many other area ranchers, including George Mayfield, who had come up from Oregon and owned the 144 Mile, and Julian Fry, an old Etonian, who ranched at Lac La Hache and became secretary of the CCA on its formation.

Hard though the times were, cattlemen entertained graciously when the occasion demanded. When the Livestock and Fair Association put on a banquet for judges and buyers at a cattle show, the table gleamed richly beneath ornate chandeliers; but despite the protestations of a visiting politician that the cattle industry was "building up the Cariboo," beef growers knew the struggle they were having. G.W. Felker, of Lac La Hache, complained to the *Williams Lake Tribune,* "Should farm and ranch land be assessed as high for taxes as when produce was two and three times the price of today? ... How do they expect the ranchers to exist with beef cattle selling at from one-and-a-half cents on the hoof?" At one ranch the owner was furious when he discovered his foreman had bought a case of eggs, and told him not to buy any more because they could not afford such luxuries.

Marketing became an increasingly important—and contentious—issue amongst the cattlemen. Some favored government involvement in a beef marketing scheme, others opposed the idea. What was clear, by the beginning of 1936, was that some form of organized approach was essential. Martin, who had been elected vice-president of the Cariboo Stockmen's Association in 1934, considered with his fellow directors a bold idea for a cooperative marketing scheme with headquarters and a sales agent in Vancouver, British Columbia. The directors agreed to sell all their cattle through their Vancouver agent and urged all members of the association to follow their example. The scheme eliminated the old approach—in effect since the inception of livestock production in B.C.—whereby the packing houses sent their buyers to the Cariboo and cattlemen sold directly to them. This method had aroused considerable resentment among ranchers, who suspected the buyers of being in collusion, and blamed the packers for the poor position in which they found themselves in the 1930s.

By the spring of 1937 the cooperative marketing scheme initiated by the Cariboo Stockmen's Association was being welcomed throughout British Columbia and farther afield as one of the most feasible plans yet developed to put the industry on its feet.

Several other responsibilities claimed Martin's attention besides running his ranch and serving as a spokesman for the cattle industry. He kept an eye on the Lodge and the general store and looked after the books for his various businesses with some help from the store manager. He was the guardian angel of the water and power systems, getting up in the middle of the night to give resuscitation when needed. He was the village postmaster for many years, and also had a mail contract—losing a few more nights' sleep as he waited for mail to arrive at Exeter station via the Pacific Great Eastern railway (known locally as "Please Go Easy," or "Past God's Endurance"). He served as the Imperial Oil agent—the company presented him with a diamond pin after twenty years' service—and built the first school in the community in 1935, giving a sub-contract to a local carpenter who worked for him on the ranch. The school was located near the PGE railway track about two miles from 100 Mile House so that the ten pupils would all have a similar distance to walk.

In April 1937 the original 100 Mile House stopping place suffered the fate of so many other old buildings in the Cariboo: it burned down. With the Lodge in place and doing a thriving business, the ancient structure was not lamented too much, except for its historical interest. As Martin remarked later, "There was a great loss of life—none of it, fortunately, human." Martin had developed acute appendicitis and driven himself down to hospital in Kamloops shortly before the fire. Edith accompanied him, along with Ben Thompson, who looked after the chores at the Lodge, and mercifully took over the driving about halfway there. While in the hospital, Martin used the time to design a new bunkhouse for the ranch crew, who had been quartered in the old stopping house. There was no doctor in the 100 Mile House area, of course, in those days, and no hospital.

Toward the end of the 1930s the Cariboo-Chilcotin was attracting many visitors from the western United States; there were more California license plates on the Cariboo Road than B.C. ones. Ex-U.S. president Herbert Hoover, of Palo Alto, California, stayed at the 100 Mile House Lodge for a night in July 1938, while on his way south with some friends following a trip to the Chilcotin. In the fall of 1938 Martin decided to make a trip to California to publicize the 100 Mile Lodge. Aurel Csanady, his wife's brother, who had come out to British Columbia from Hungary the previous year and was staying at the Lodge, accompanied him. They did some promotional work in Portland and then drove down to Los Angeles, where Jimmy McLarnin, a Vancouver boxer who had won the world welterweight title in 1934, introduced them to celebrities like Bob Hope, Bing Crosby, Martha Raye, and Oliver Hardy. However, while Martin was well received, the trip did not bring an influx of movie stars to the Cariboo.

During the summer of 1939 Martin met a kindred spirit when an urbane Englishman by the name of Conrad O'Brien-ffrench arrived at 100 Mile House to visit Martin's sister, Romayne, who was staying at "the 100" at the time. An immediate bond developed between the two men.

Born in London in 1893, the son of an old Irish family, O'Brien-ffrench had sailed for Canada at the age of seventeen to join the Mounties. Later, captured by the Germans at the battle of Mons in the First World War, he gathered valuable classified information from other prisoners and passed it on to British Intelligence in London by means of invisible ink. So began a career as a secret agent that is said to have inspired his friend, Ian Fleming, with the idea for James Bond. Under the guise of playboy and sportsman, O'Brien-ffrench spied on Hitler's Germany throughout the 1930s, sending information to Stewart Menzies, then head of British Intelligence and supposedly the model for Ian Fleming's "M".

In March 1938 O'Brien-ffrench broke his cover by phoning London with news that German forces were advancing into Austria. After making a narrow escape into Switzerland, he resigned from the secret service and decided to find a new home in Canada. He was looking into purchasing property in British Columbia when he met Romayne Cecil during a skiing holiday in Switzerland. When Lady Romayne told him that she planned to visit her brother in British Columbia in the summer of 1939, he was immediately interested. "I might come out and see you," he said. "I'm planning to go to British Columbia myself."

So it was that on a warm summer's day in June 1939, as the war clouds massed over Europe once again, Conrad O'Brien-ffrench drove to 100 Mile House to keep his rendezvous with Lady Romayne. In his autobiography, *Delicate Mission*, he described his first meeting with Martin:

On reaching the floor of the valley where the road crosses a stream I found a lodge, a barn, and some corrals. Tired and hungry, I wondered what sort of a welcome I would receive. Close to where I stopped I could see a muskrat swimming in the stream and not far from him a cinnamon duck rested on the bank. For a while there was no other sign of life. I watched a kingbird making sallies from a fencepost. Then a screendoor opened and I saw Lord Martin Cecil coming out to meet me. What seemed to be an unimportant encounter was to prove the crucial step into a period of my life which brought to light new and greater purpose in living. Martin stood before me; in his regard was depth of sincerity, perhaps a little shy, for he was neither cold nor effusive, neither was he talkative nor taciturn, gushing nor indifferent, but there was a quality which touched me immediately.

At first I was on my guard, but instinctively I knew that he had stimulated a responsive chord in me. Had I evaluated him socially as a Cecil, of noble parentage, I would have lost him wholly. As a rancher, too, he was well thought of throughout the West, yet ranching pure and simple gave me no focus on this man. He gave me a sincere handshake and the friendliest of smiles which evaporated any doubts I might have had about my welcome, and introduced me to his wife. Thus, in a few moments, was a new beginning made.

After a few days O'Brien-ffrench left to return to his home

on Vancouver Island. But his close friendship and agreement with Martin would continue until his death in 1986.

In August 1939 the Victoria *Colonist* announced the death of C.G. Cowan. Cowan had been in poor health for some years, and was living in a nursing home in Victoria. The world was now moving inexorably closer to the brink of another world war. Germany sent U-boats on patrol into the Atlantic and deployed the pocket battleships *Graf Spee* and *Deutschland*. On August 30 German SS troops in Polish uniforms attacked a German radio station at Gleiwitz to convince the world that Poland was the aggressor, and on September 1, without declaring war, Germany invaded Poland. Both Britain and France sent ultimatums to Hitler to withdraw, but there was no response. On September 3 Neville Chamberlain announced that war had begun. Lady Romayne hastily left 100 Mile House and returned to England.

By no means was Canada unified in the decision to join in the conflict. But while several French-Canadian MPs spoke against the war, Mackenzie King's pledge that there would be no conscription satisfied them sufficiently that they did not force a vote on the issue. As the nation girded up for an all-out military effort, the Depression, which had already begun to lift in 1937 and 1938, finally dissolved. There was work for everyone now—in the armed services, in factories and munition plants, and on farms.

During the pre-war period Martin had been gradually awakening to a deeper sense of purpose in life. Early in 1938—as he struggled to keep his affairs in order in the face of increasing stresses and strains—he began to sense a source of wisdom and peace within himself that was always there no matter what was happening around him. As he wrote in a letter a year or two later, this first touching of a transcendent state in himself brought "such happiness and peace as I had never known before, although outwardly my affairs and my life were in something of a turmoil." Martin's dawning interest in the spiritual life was shared by Edith, who had been finding it increasingly difficult to adjust to her new life in the Cariboo, and suffered from periods of depression. She and Martin enrolled for a while in a course called "The Impersonal Way," and read books on Rosicrucianism and related subjects. A favorite work of Martin's was *The Prophet,* by Kahlil Gibran.

One day, amid all the other hectic events that transpired in the fall of 1939, a friend of Edith's who lived in Vancouver told Martin about an American visionary named Lloyd Arthur Meeker. The woman suggested Martin write and ask to be put on Meeker's mailing list, which he did. On December 23, 1939, as the conflagration of World War Two intensified, Martin received a letter and a parcel of literature from Meeker.

It was a pivotal moment for Martin. He realized, as he went through the material, that he had been starving for this rational, common-sense approach to spirituality which answered all his questions. But not until the spiritual food was "put on the table" did he realize just how much he had been starving. He devoured the literature—hardly sleeping at nights. On December 28, 1939, he wrote his first letter to Meeker. "It was indeed a great joy to receive and peruse your correspondence and literature," he wrote, "adding to my conviction of the presence within of One who surely knows. I recognize how necessary it is to cut a channel so that reality may flow out into expression through every activity of my life. I am thankful to have been led to you, and I give thanks for the many truths you have sent forth that have already found a response in me."

The stage was set for him to meet Meeker in person.

11.

LLOYD MEEKER

While Martin was taking part in a children's fancy dress party at Burghley House in his white owlet costume made of velvet and trimmed with green ostrich feathers, or studying his arithmetic and piano with "Miss Win," Lloyd Arthur Meeker, just two years his senior, was grubbing sagebrush all day under a burning sun in the Colorado wilderness.

Born in 1907 in Ferguson, Iowa, the son of a poor farmer and circuit minister, Lloyd Meeker spent his boyhood on the family homestead in the foothills of the Rocky Mountains southwest of Grand Junction. It was a lonely, barren land dotted with pinion trees, juniper trees, and sagebrush. From the age of seven, Lloyd and a younger brother worked in the fields, clearing the ground for dryland farming under the watchful eye of an enormous Indian squaw. The boys walked three miles each way to a little schoolhouse, with instructions to return home as soon as school was finished to work on the farm; their only chance to play with other children was during the noon hour at school. They lived in a tiny shack, subject to a stern father who sought to impose his religious beliefs through harsh discipline. During the flu epidemic of 1919, while he himself was sick in bed, Lloyd heard his mother, his only solace, breathe her last breath.

After years of hardship and privation—including a forty-day bout of typhoid fever that nearly took his life—Lloyd Meeker left home at the age of sixteen. In reaction, mainly, to his father, he had decided that there was nothing to religion, and he considered himself an atheist. He rode the freight trains and eventually, in Fort Scott, Kansas, found work as a laborer on a building construction gang, unloading carloads of cement and plaster on stifling hot August days, and manhandling steel when it was so cold that hands and gloves stuck to the metal. By 1929 he had worked his way up to become office manager for the building company. Then came the stock market crash: without warning he lost his job and his house; his bank closed its doors; he was left with nothing.

It is difficult to imagine what Meeker went through during the early years of his life. I do know, however, that thrust into extreme, even life-threatening situations—such as the camps of the Gulag—men and women sometimes discover a source of strength within that amazes both themselves and others. It is as if the intensity of their misfortune opens the door to a transcendent experience hitherto unknown and undreamed-of. Mihailo Mihailov, Yugoslav writer and activist, wrote in a Belgrade prison, "When a man has got rid of all that ties him, a mysterious thing happens to this outwardly unfree, but inwardly at last utterly free person. In the depths of his soul there rises up a mighty force, which not only endows his exhausted body with incredible powers of resistance, but, in strange ways which we do not yet fully understand, also begins to affect the visible world ... to determine events over which—I repeat once more—he can so far as we know today have no influence, but which become his salvation." Solzhenitsyn wrote of a strange inner warmth that seemed to come from another world and save a person from freezing in glacier ice. Such accounts come not from abstract thought, but from personal experience of this inner force.

Although the context was different, perhaps something of this nature happened to Meeker. Certainly, as he went through a process of relinquishing external ties, he too found himself "coming free" internally. Since leaving home he had read many books on philosophy and psychology, seeking answers to life's age-old questions—seeking, also, for ways to improve his own health, for he had suffered much in his young life and had often worked despite severe physical discomfort. But the more he read other people's writings and other people's philosophies the more he began to sense that fundamentally the answers to his questions were all present within himself.

Arriving in Nashville, Tennessee, Meeker made a fateful decision: to ignore all he had read, all the things he thought he knew, and place his complete trust in this sensing of a spiritual source, a spiritual authority, within himself.

During a three-day period that began on the morning of September 13, 1932, his initial sensing in this regard underwent a profound metamorphosis. He found himself ushered, as it were, into a vivid recognition of humankind's true identity in spirit. As the three days progressed—accompanied by a depth of peace unlike anything he had ever known before—he eagerly wrote down the thoughts and realizations that flooded into his consciousness. When he read over what he had written, he found the answers to many of his questions. Toward the close of the third day, September 15, he knew that the experience he had had would not return, and it never did. As he would remark later, there was no reason that it should.

He was a young man of just twenty-five. But by September 16, 1932, Meeker had determined to give the rest of his life to offering his new perspective and understanding to others. He had no material resources; he was unknown except to a few people in Nashville; but he felt confident that whatever was needed to fulfill his commission would always be provided. He began to devote himself to healing, and as his work became known, more people began to call upon him for physical and spiritual help. During the 1930s and early 1940s he traveled extensively throughout the United States and Canada, lecturing and teaching, using the pen-name Uranda. Considerable interest developed in California, and he gave talks and classes in such places as Oakland, San Francisco, and Long Beach, where he had an office for a while. So that he could operate more freely and effectively in the world as it was, he incorporated his program as a church in January 1940, under the name Emissaries of Divine Light.

What message did Meeker bring, that influenced Martin, and others, so powerfully? It was the same truth that Jesus, Buddha, and other teachers have brought—the only truth there need be—namely the presence within each person of light. Each one may say, "I am that light." It may be true that special human beings have emerged from time to time to remind humankind of the universal truth they have forgotten. But that does not mean that everyone else is condemned to some pale, secondhand experience. Each person has unique potential. It would appear, in considering what is called Christianity, for instance, that the real point which Jesus brought has been overlooked or forgotten. "Follow me," he said. Do what I have done. That is the word of any true master, for ultimately there is no profit merely in looking to someone else or trying to believe someone else. "I am the light of the world." Meeker reaffirmed and restated in his own unique, dynamic way the opportunity that is present for anyone to find identity with spirit, with light—to BE that, because that is what one chooses to express.

Incarnate within all people, Meeker affirmed, dwells an aspect of the spirit of God, eternal and perfect, having no need to evolve or to grow. Where there is a willingness to align with that, so that mind and emotions become clear channels through which the essences of that incarnate spirit may find release, all problems immediately begin to dissolve. Those problems only appeared in the first place because of misalignment—because of a lack of the experience of union with spirit. This union, said Meeker, is love—perfect love. In a booklet entitled "Seven Steps to the Temple of Light," he stated:

Love is the law of expression in every human being, and he who harmonizes with this great law is he who begins to live a glorious and happy life, here and now. There is only one way whereby anyone may harmonize with this law, and in this one way is certain and absolute success, no matter who the person is, no matter what he has or has not done, and no matter where he or she is or is not. This way works for the bond and the free, for the poor and for the rich; for it is no respecter of persons. It cannot be purchased with money, nor can worldly position or fame secure it. A college education is no help in finding it, and lack of so-called education is no hindrance, for it is freely available to all.

To walk in this one way, said Meeker, requires a willingness to let go—to stop struggling with one's problems and difficulties. He identified the attitude of thankfulness as an essential requirement. It is not so much a matter of trying to be thankful for difficult or unpleasant circumstances. But unless there is a fundamental thankfulness for the gift of life, the gift of spirit, however this may be perceived, there is no chance for the wisdom of spirit to find release and bring creative change, he declared.

Martin, whose own family motto proclaims the importance of "One Way," had been deeply touched by his first exposure to Meeker's literature. So he was excited—and nervous too—when the opportunity arose to meet the man in the flesh. Martin sometimes rented a house in Vancouver for part of the winter to escape the long freeze at 100 Mile House. He and Edith, with their four-year-old son, Michael, were staying at a home on Blenheim Street when Meeker wrote to say that he proposed visiting Vancouver in April 1940 to give a series of talks. Would Martin help to make the arrangements? "Yes," said Martin, and proceeded to find a suitable room in the old Georgia Street Medical Building where Meeker could speak.

On a pleasant spring afternoon Lloyd Meeker drove to the house on Blenheim St. to meet Martin and Edith and discuss arrangements for the meetings. They sat down in the living room while Edith made some tea. Michael was present, and at one point Meeker took him up and held him on his knee. Michael remembers his piercing look to this day. "It was as if he looked right through to the core of my being," he recalls.

For Martin, his first meeting with the man who had brought such a new and exciting dimension into his life had a natural and satisfying feel to it. Martin didn't ask any great philosophical questions—it would have been out of character—but was concerned to check over the plans he had made to make sure they were suitable. He was, of course, exercising his powers of observation to the full. Following the visit he wrote: "Meeker arrived. I was extremely nervous, as he well knew, and concerned as to what manner of man he should be. So many people have said that at first they were disappointed in him. I don't think that was my first reaction. It seemed more like relief. In his unassuming charm he revealed reality in a manner which appealed to me as genuine and fulfilled my expectations to the greatest possible degree. After that first meeting all my doubts and fears vanished away ... "

One thing that was a surprise was Meeker's youth. Considering the wisdom and common sense revealed in his literature, Martin had expected him to be older, whereas he looked about the same age as Martin himself—which he was. What puzzled Martin most, though, was his feeling that he had somehow met the American before. There was a sense of closeness and familiarity which he was at a loss to explain.

Martin was particularly moved by Meeker's voice. It had a rich, magnetic quality which stirred him deeply, especially when Meeker was giving his public talks. The talks—attended by Edith's brother, Aurel, and Conrad O'Brien-ffrench, amongst others—prompted Martin to write to his new friend on April 18: "You have given me a picture of true greatness, a picture that has been hovering in my consciousness but had never until now become fully revealed to my outer vision. It is a picture that is joyously familiar now that it is more fully perceived."

12.

TOUCHSTONE

It does take a little time to relinquish old habits and attitudes and bring one's thoughts and behavior into alignment with the real character of spirit. A process is involved. By the way he lived, the easy but strong way in which he received and welcomed everything that life brought to him, Lloyd Meeker showed Martin how the process works: how it is possible to act and express in the flow of the spirit continuously. He was a touchstone, exemplifying the truth which was inherent in Martin too, but which Martin did not yet know because he was not giving it adequate expression.

As part of this process of re-alignment, or spiritual maturing, Martin maintained a consistent correspondence with his guide and mentor from 1939 onward, sharing his thoughts and realizations on life, and happenings in his world. One area which caused him considerable anxiety and soul-searching in the early days of his association with Meeker related to the necessity of letting go of some of his hereditary ties. This was emphasized in the beginning stages of World War Two when Martin faced the question of obligation to "King and Country." He wrote to Meeker:

Originally serving in the British Navy it is truly wonderful how all direct hold upon my services has been released. However being born a member of an aristocratic English family, a certain friction has developed between them and myself. I know that some important and necessary changes are taking place through this, because probably what troubles me most is the feeling that I am losing something. This 'something' I must willingly let go. Though my immediate course is clear to me and I cannot falter, yet the dropping away of these ties causes a certain uneasiness within. May the spirit of God, to which I can alone be loyal, guide me in the way.

Later in 1940 Martin wrote poignantly of a small green shoot which he had observed pushing its way up through a well-worn pathway near the Lodge: "Shortly the cracks widened until finally a dandelion plant came forth and as it came the hard earth which had lately been its prison broke up and fell away, releasing the plant out of the darkness of the earth into the freedom of the air, to manifest its true nature in fullness, reflecting the radiant sunlight in the beauty of its golden flower. So may it be for a human being, as this little plant reveals. Why are we disturbed when the earth begins to crack?—for thereby is our release made possible."

The changes happening in Martin affected his view of his work, both the managing of the ranch and his work with the cattle industry. Before, he had seen his activities simply as tasks that needed to be done, but now he began to recognize the larger implications that lay back of them. He struggled to clarify his new awareness: "I have the impression that that which I have hitherto looked upon as my 'work,' my 'job,' actually only provides the opportunity for my true service," he wrote at one point. "It is as though a field of service is just coming within the range of my vision which as it were runs parallel with my material 'job.' This perhaps is rather confused, naturally I suppose because it is still not clear to me, but I rejoice to release it as it is, in the knowledge that greater understanding will come."

Martin's responsibilities in the cattle industry increased substantially during the war years. He became president of the Cariboo Stockmen's Association in 1940, and three years later played a key role in forming a new organization, the Cariboo Cattlemen's Association. This was part of a move to unify the cattle industry in British Columbia and also to cooperate more fully with the provincial B.C. Beef Cattle Growers' Association and the Alberta and Saskatchewan cattle associations; steps were also being taken to form a Canadian Cattlemen's Association. Addressing the inaugural meeting of the Cariboo Cattlemen's Association in Williams Lake in 1943, Martin, the CCA's first president, noted the movement toward greater unification in the industry. "However it is evident that new constitutions are in themselves not enough," he went on. "In time of war we wake to the realization that national accomplishment is based upon individual effort and initiative—the same principle is true with respect to our lesser activities here."

In September 1944 Martin and Edith attended a two-week summer session which Lloyd Meeker convened at a ranch near the town of Loveland, Colorado, where he had now established his headquarters. This was an important event for Meeker, bringing together various key people who had been drawn to participate in his program. Among those who attended the summer session were Richard and Dorothy Thompson, and two sisters from Ravenden, Arkansas, named Kathy and Rosalind Groves. Kathy had had her first introduction to the Emissary program at the age of twelve, when she enrolled in the "Young Explorers' Club," as it was called; even at that tender age, her letters acknowledging the club material evinced a remarkable quality of understanding. She met Meeker for the first time in the summer of 1941, when with her sister Billie she attended a talk which he gave in Pittsburgh, Pennsylvania; she was then only fifteen, but fell deeply in love with him. Following the 1944 summer session, on September 30, Meeker hosted a party for Kathy on her nineteenth birthday, and they were married a few months later. Kathy's sense of humor, brightness of spirit, composure, and understanding were all essential in the development of the Emissary program in the years following. She had a lovely voice and delighted to sing both sacred and secular songs; she was, Meeker once declared, "a woman without guile."

For Martin, one of the highlights of the summer session was meeting other Emissaries—he discovered it was as if he had "known them all his life." He also noted with delight some improvement in Edith as she relaxed into the current of love that was present. She had been under emotional and physical stress for some time, and he had been doing his best to help her into a new, more positive approach to life, though aware at the same time of his own remaining restrictions and limitations.

For many years Meeker had dreamed of acquiring a permanent headquarters where men and women might come together and discover how to align with the pulse of spirit, and how to express that spirit collectively. Toward the close of 1945 the dream became reality. Out riding her horse one day in a valley northeast of Loveland, Rosalind Groves came upon a rundown, 123-acre dryland farm with a "for sale" sign hanging forlornly from a stake by the road. Meeker and his little band bought the property and named it Sunrise Ranch. A less promising place would be hard to imagine. The land was arid and desolate. There was no water or electricity. The few buildings—an old ranchhouse, a smokehouse, a large barn and granary and one other small house—were ramshackle. But Meeker had a vision of what could occur. With his new bride, Kathy, he moved into a room in the ranchhouse, and on December 25, 1945, a small group sat down to their first Christmas dinner there. Billie Cable—sister of Kathy and Rosalind—recalls the event: "The kitchen walls and ceiling were covered with coal dust and smoke accumulated over the years, but the house was soon cleaned and made ready. As we prepared the meal Kathy lovingly set the table with a white cloth and arranged a centerpiece of evergreen boughs and bayberry candles. The food was blessed and enjoyed by all, and afterward we gathered in the living room to share a period of meditation. In this atmosphere of joy and solemnity we offered our agreement and support in the new undertaking." About fifteen people shared in the beginnings of Sunrise Ranch. Meeker attended countless auctions, buying essential machinery and livestock, including a pair of mules which were used for the first plowing of the land.

The friendship and love being nurtured by this small group of people in the foothills of the Rocky Mountains contrasted sharply with the fear and hostility sweeping the larger body of humankind. Hopes that the Allied victory in the Second World War would usher in an era of peace and brotherhood were turning to ashes as the so-called "iron curtain" descended across Europe. Small though the numbers were, however, the seed which had been planted at Sunrise Ranch had greater significance than might have met the eye: for it takes only a few people in agreement with spirit to let the power of spirit be brought to bear in the world.

13.

TURNING POINT

In his book, *The Healing of the Planet Earth,* Alan Cohen writes: "Every time I have made a step forward in my personal and spiritual evolution in the areas of health, success, or relationships, I have had the feeling that I was about to be punished, die, or lose something. When I found the courage to go ahead despite the ego's gloomy counsel, I discovered deep peace, greater aliveness, and immeasurable gain."

The ego, as Cohen terms it, does indeed put up a challenge if we begin to move in a new direction in life. A quality of spiritual determination is essential if we are to overcome that challenge and discover that greater experience of ourselves which is our birthright and our true destiny. Lloyd Meeker described what is required in an address later entitled, "Through to Victory":

When it appears that there is a wall in front of you, and perhaps you can begin to see it coming up, as it were, learn to keep right on going. Learn to walk right up—TO the wall? No. If you walked right up to the wall, how close would you get to the wall? A hundred yards, or ten yards, or six feet? You walk up close to the wall, and stop, and wait for the wall to vanish? It does not work that way.

You learn to walk not up to the walls, but right through them. You keep right on going. And you will find that when you get right up to it, pressing against it, it opens. Or you find, suddenly, that it just looked like there was a wall. Anyone round about would have sworn there was a wall. But when you get up there close enough, it begins to move back.

Don't be deluded. Keep right on going, and the wall will open, or prove to be just a mirage. Perhaps it isn't even there, but just looks like it. Hold it steady and keep going.

This is the only way to victory. If walls can stop you, you will be stopped. If human beings can upset you, you will be upset. If discouragement can turn you aside, you will be turned aside. That is the situation. It is only for those who refuse to stop for any wall, who keep right on going, that there can be any true sharing in the victory.

A large and foreboding wall loomed before Martin in 1946, when through a series of events he faced the prospect of losing those who were most dear and close to him.

One such loss involved his son Michael. Since Martin's brother, David, did not have a son, Michael would one day succeed to the Exeter title. Partly because of this, Lord and Lady Exeter wanted Michael to come to England to complete his education and become familiar with the social setting there.

In the late spring of 1946 Lord Exeter visited 100 Mile House. One reason for his coming—it quickly became apparent—was to suggest that Michael be allowed to come to England that fall and attend a private preparatory school before going to Eton College.

Martin faced a dilemma—the prospect of losing his only son during a tender and formative period in life. Michael might not even return to Canada, might become enmeshed in the traditions and values which Martin himself had left behind.

But as he thought about the matter and wrestled with the rather few options open to him, Martin knew that it was the right thing to let Michael go to England. Apart from anything else, he was not in position to give the boy the care and attention that he needed. So that was one heartache. Even more devastating, however, was Edith's situation. Brought up as she was in a sheltered European background, and being by nature a sensitive, delicate woman, Edith had simply not been able to adjust to her new life in Canada. It was as if she were a lovely flower who could not survive being transplanted to the rugged soil of the Cariboo. A tendency to depression had taken its toll both physically and emotionally, so that despite occasional upturns—she had benefited greatly from the summer session of 1944—she had been a virtual invalid for some years. By June 1946 it was clear that she needed special treatment, and Martin decided it was necessary to admit her to a sanatorium in Ontario.

It was painful and wrenching to have to leave Edith all on her own in the East and return home without her. In a letter to Meeker written in June 1946 Martin described the trip as "perhaps the most nerve-wracking experience I ever went through." Air travel was primitive and unreliable in those early postwar days, and he was delayed in Toronto waiting for a flight back to British Columbia. While he was walking down a Toronto street, a cinder blew into his eye. It was a Saturday, and the doctor at the hotel where he was staying was off duty. There was a nurse, but she was reluctant to do anything for fear of damaging his eye. With the pain of the cinder and his emotional anguish he was unable to sleep that night. After tossing and turning for a while he rose from bed and decided to write something: partly he wanted to distract himself, but more than that he knew he had to meet the grief and despair seething inside him. Lloyd Meeker had introduced him to a way of love and truth, shown him how it worked, but as Meeker often pointed out, everyone has to prove that way for him- or herself. "It takes a heap of living," he would say, in his resonant voice, "to prove the truth."

As Martin's life reached a point of intense crisis in an outer sense, he let go to the inner reality that he knew was of supreme importance. Alone in his room in the Royal York Hotel, he wrote a poem which he later entitled "The Abyss of Nothingness." He did not think of himself as a poet. But because the words came from his heart they brought peace, a renewed sense of clarity and direction.

> There are times in life when the human heart
> Is thrust through, it seems, by the fiery dart
> Of remorse and pain and a wordless grief,
> So that pity of self steals in like a thief.
>
> Then what shall we do when we reach the place
> Where courage is gone and it's hard to face
> The years ahead? Shall we quit?

Or cry? Shall we weakly submit
To the voice of discouragement whispering low:
'Why should I, of all people, be treated so?'

And then perhaps sad conscience brings
Its burden, and there rings
Through troubled mind a tragic air
Of endless discord and despair.

'O God, where art thou in my desolation?
I feel thee not, nor hope for thy salvation.
But thou art near at hand, my Lord, I know,
Because there have been times I knew it so.
Brought low am I; my strength is gone;
What is there left to lean upon?'

In the silence of exhaustion the fevered mind is still,
And there comes a relaxation of distraught human will.
So through the darkness of the night the first faint
   flush of dawn
Comes gently, giving promise of a new day to be born.

'O Man, why fear? Why doubt ye me?
Who am the source of life in thee.
Dost live and breathe? Canst see and feel and hear?
How else these things except your Lord be near?'

So sweetly as the summer breeze
Stirs soft response in grass and trees,
The silent sounding of thy word,
Like distant music faintly heard,
Awakes an echo in my heart

To bring assurance that thou art
The One whose presence now in me
To life gives meaning, sets me free
From doubt and fear and bleak despair!
For in thy strength I surely dare
To let all burdens roll away
And give to thee the fullest sway
In me. For I am not mine own,
Nor any man's, but thine alone.

This earthly form which I have thought as mine,
With all its weakness and its woe, is thine.
It take, O Lord, and train me in thy way,
To let thy life be lived in me today
And always.

He had met the challenge of his own human nature—the ego's "gloomy counsel"—and emerged victorious, as Arjuna did in the Hindu sacred text, the *Song Celestial*. Everything he looked at now he saw from an entirely new perspective. He had proven for himself the truth of Krishna's beautiful words to Arjuna: "Who doeth all for Me; who findeth Me in all; adoreth always; loveth all which I have made, and Me, for Love's sole end, that man, Arjuna! unto Me doth wend." It was as if he had climbed to the peak of a mountain and could see clearly all around him. Before he set foot on the mountain—while he was still a plains-dweller—the mountain had seemed majestic but remote, as does God in the view of many people. Now, that had changed. He was standing on the mountain. He was part of God, and God was part of him.

Standing on that higher ground he became aware of a spiritual commission as bright and expansive as a Cariboo dawn. He shared his realizations in a letter to Meeker in February 1947. He spoke of the one universal spirit of love, truth, and life which he had come to know. He spoke of the magical process by which a body of men and women was being brought together to provide a means for the action of that spirit on earth. But most importantly he acknowledged the responsibility which he now knew was his to assist in that drawing together—to assist all who were willing into a greater experience of the unity and love which is humanity's true destiny.

Meeker received the letter with deep thankfulness and delight. For several years he had offered all the assistance and encouragement he could to both Martin and Edith, seeking to inspire them to experience for themselves the victory that is inherent in spirit. That victory—with its various components, such as strength, wisdom, and understanding—cannot be imposed; it can only be received. Of course, a person has to be willing to let the necessary changes come so that it *can* be received. It was evident that Martin had been willing to do that, and that he stood now with Meeker in complete and absolute agreement. Within a few days, the latter sent a personal message to those on his mailing list announcing that he would hold a special gathering to acknowledge the spiritual leadership which Martin was providing.

The acknowledgment took place on March 23, 1947, at a house near Riverside, California, where the Meekers were staying. Martin drove down with Conrad O'Brien-ffrench; the two slept on cots in a hayshed among some farm machinery, waking up one morning to find a cow breathing in their faces. But if the accommodation was sparse and rustic, there was no mistaking the quality and richness of spirit that was present. It was a time of great gladness, born of the recognition that the agreement between Lloyd Meeker and Martin would provide a base for an expansion of the Emissary movement, and allow the invitation of spirit to be made known to larger numbers of people.

As the ceremony proceeded Martin was keenly aware that he was not being appointed to a position. Meeker was merely acknowledging a fact—the fact that at the age of thirty-seven, Martin had come into position to fulfill a responsibility of great significance for his world. His famous ancestor, William Cecil, was thirty-eight when Queen Elizabeth 1 appointed him her principal Secretary of State.

Considering the enormous difference in their backgrounds, I have often thought how remarkable it was that Martin and Lloyd Meeker met at all. Who would have expected it: one born into the Old World, one into the New; one born to privilege and wealth, one to poverty and labor; with an ocean between. But meet they did. And the way in which it happened, the way in which so many seemingly coincidental factors sorted themselves out and intertwined, proclaims loudly to me that there is a destiny, a right and true destiny, for each one of us, and if we open and let go to the impulse of spirit we find that destiny; we find the one—or ones—whom we need to find.

Precisely because of their great differences, of course, the friendship of these two men covered the whole spectrum of humanity, from the rich to the poor, the privileged to the dispossessed. It is noteworthy also that Martin, although an aristocrat to his fingertips, trained all his life to lead, accepted

Meeker's authority from the beginning. It reminds me of the Roman centurion who accepted Jesus in the same immediate, unquestioning manner. I never met Meeker, and thus never saw the two of them together, but I have spoken with many who did, and they all affirm Martin's unwavering respect and love for his mentor. There was never any question, for Martin, where the authority lay. If Meeker was speaking, at a public gathering perhaps, or to a smaller group, Martin would sit quietly, listening intently. Only when he felt it appropriate to complement the other's words, or offer support, would he speak. But although there was a mutual respect between the two men there was a lightness too; they enjoyed each other's company; a current of fun flowed between them that could be as lively and sparkling as a mountain stream.

I suppose this all shows that there is a spiritual design, or aristocracy, which for those who are sensitive to it, supersedes earthly status or position. It took a while, but the British did, finally, acknowledge a certain spiritual quality, or nobility, in Gandhi. Francis of Assisi earned the adoration of rich as well as poor people.

As the work which Lloyd Meeker and Martin shared continued to expand, something began to come within the range of human experience on earth which transcended in every way the boundaries of orthodox religion. Some of Martin's family found this highly disturbing. After all, the Exeters had upheld the Church of England ever since William Cecil and his Queen established the Protestant movement in Britain three centuries before. It looked as if Martin was deviating from this tradition—from the cause of Christianity itself. But he wasn't really. Having awakened to his own spirituality, he was simply responding to the imperative for change and clarification that is part of life's very nature; the Church of England had been born out of this same imperative.

The image of an awakening is a metaphor which pervades many of the early myths and legends of the human race. King Arthur sleeps, waiting for his country's call. In the Volsung Saga, Brynhild waits in a deserted castle surrounded by a barrier of fire to be awakened by Sigurd. In the story of Sleeping Beauty, the brave prince ignores the warnings of those around him, plunges through the dark and dangerous forest, and comes to the castle where the beautiful princess lies. "Is it you, my prince?" she asks, opening her eyes. "I have waited for you very long."

It may be a fairy tale, but it illustrates something. We all grow up amid a tangle of opinions, customs, and beliefs, and sometimes they are so thick and firmly established as to appear impenetrable. Yet if we have the courage to wield it, the sword of truth will cut through the snarl and bring wakefulness.

14.

SPIRITUAL PIONEERING

In the spring of 1948 Martin initiated a small communal group at the Lodge, based upon the same precepts as the community which Lloyd Meeker had established at Sunrise Ranch. This too would be a place where people might learn how to live together in the fulfillment of life's purposes.

Usually, when people start an intentional community, they choose a place that—like Sunrise Ranch itself, or Twin Oaks in Virginia, or the well-known Findhorn community in Scotland—is secluded, tucked away from the hustle and bustle of everyday existence. Such was also the case with Brook Farm, a pioneer venture established in New England in 1841 by the Transcendentalists; while it was short-lived, this important social experiment influenced Emerson, Thoreau, Whitman, the publisher Horace Greeley, and many other prominent thinkers of the time.

The Lodge community, at its beginning, nestled amid fields, trees, and a stream—all the things you might think of as needful to such an enterprise—with scarcely a building or person in sight. The picture changed dramatically, however, as 100 Mile House boomed, becoming what is today a modern municipality of some two thousand people, serving a trading area of fifteen to twenty thousand. This has given a unique flavor and character to the Lodge community. Far from being secluded, it interacts with the larger village on a daily basis and in many diverse ways.

Lumber was the catalyst that transformed 100 Mile House and earned it the distinction, during the 1950s and 1960s, of being the "fastest-growing village in British Columbia." Martin had always felt that 100 Mile had the potential to be a larger center, with its strategic location on the main north-south highway, its proximity to a railway, and the fact that it is a junction for a network of roads leading to outlying areas. This was one reason he had established essential services such as water and power—so that there would be a base for future development. When the demand for timber reached into the interior of British Columbia following World War Two, it was no surprise that 100 Mile House should become a hub for the industry.

The initial thrust in this larger development came when two brothers from Saskatchewan, Rudy and Slim Jens, set up a planer mill near 100 Mile House in 1948. Not long afterwards they approached Martin about the possibility of building some houses in 100 Mile House for themselves and some key employees, and so it was that a modern-day town began to sprout upon one of Martin's best hayfields. Anxious not to lose his land forever, should the lumber boom prove to be short-lived, he developed a scheme whereby he leased his land for residential or commercial use for an initial twenty-five-year period.

One of the major benefits of the lease approach, as it worked out, was that it enabled Martin to keep a hand on the development of the village. Rather than springing up in a sprawling, haphazard way, 100 Mile House carried the imprint of his own orderly nature from the beginning. He hired a land surveyor. As new businesses developed, he made sure they were not hampered by undue competition.

Playing an essential part in all this was Ross Marks, a young man with a ready grin and a thick swatch of auburn hair, who came out to assist Martin during his summer holidays in 1948—and never did return home. A student at the Guelph Agricultural College, Marks became Martin's right-hand man, taking an increasingly responsible role in his business affairs. He also learned to maintain, and revive when necessary, the water and power systems, a most welcome and valuable contribution.

Half-a-dozen people shared in the beginnings of the Lodge community. They would gather with Martin four times a week in his living room to share a period of meditation, during which he would generally read a transcript of one of Meeker's talks. Meeker now spent most of his time at Sunrise Ranch, but traveled regularly throughout many parts of North America. His mailing list included people in California, the Midwest, and various Eastern states, as well as Canada.

Besides Marks, the Lodge family included Lena Lucyk, who worked as a waitress and cook, and Jessie Stuckey, who managed the Lodge. They were warm, irrepressible women. When a guest who was checking in asked if there were showers, Stuckey looked out of the window and replied with a chuckle, "Yes, lots, I see one coming now." Once, some city folks from Vancouver complained about their steaks. Lucyk, who had been doing her best to keep everyone happy during a long evening, sighed and forgot all about being thankful. "Would you like me to get a hatchet and chop them up for you?" she asked sweetly. At least there were no more complaints that evening.

The day began at six for the kitchen staff. Lena Lucyk would put on a robe and slippers and hurry downstairs to fire up "Leo," the kitchen range, and put the kettles on. After she had dressed, it would be time to brew the coffee and cook some cereal and bacon for the cowboys and guests. Duties for the day were written on a schedule. Someone would wash the dishes piled neatly by the sink in response to the sign, "Please stack dishes. Thank you." Other helpers would do laundry, make cottage cheese and butter, or tidy rooms.

Martin soon found that providing spiritual leadership even for such a small number was a major challenge, requiring constant care and vigilance. As the participants passed through their first year together all kinds of difficulties and potential setbacks appeared. In August 1948 Martin wrote to Lloyd Meeker: "I feel there is a surprising freedom in our function here from 'gossip.' However while there may be a goodly degree of control in this sense, it is evident that there are still divisions and reactions one toward another." Noting how two women in the group enjoyed each other's company but tended to exclude another woman, he commented: "I have pointed out that joyous friendships one with another are wonderful provided the balance is maintained; but if they come into being at the expense of the group as a whole, they are of little value." Another problem concerned a man who had come to the Lodge from Winnipeg after meeting

Meeker. He was very reserved, which made it difficult for him to blend in with the rest of the group. Martin became increasingly concerned for the individual, and one day arranged to drive him to Williams Lake—ostensibly for a haircut, but really so that they might have a chance to talk together more intimately. The ploy worked. "I am happy to report that he has begun to open up in an excellent manner," reported Martin delightedly. "He has great potentialities, and is, I feel, beginning to relax and let go of his feelings of fear and guilt."

However, no sooner was one difficulty resolved—one person helped along the way—than more problems arose, often in the area of relationships, or health. In handling all the various exigencies, Martin frequently took refuge in some humor, as when he wrote to Meeker: ". . . she continues to have great difficulty with her feet. I think it is largely by reason of the fact that she carries the weight of the world on her back that they give out." While Ross Marks' youth sometimes ran to over-exuberance, Martin found his steadiness and sense of fun a most helpful influence in all that was happening.

Toward the end of October 1948 a serious situation developed when a member of the group suffered a nervous breakdown and had to be admitted to hospital in Vancouver. She recovered well and returned to the Lodge, but the event cast a shadow over the fledgling community. On December 24, 1948, Martin wrote ruefully to Meeker, "The various stresses and strains have been telling upon my physical health to some extent. Several weeks ago I injured my knee rather painfully, and it still necessitates care. Also for a while I suffered from an aching feeling of constriction around the heart, but this has pretty well cleared now. When I consider all your years of unremitting ministry, and what you are doing there at Sunrise with fifty Emissaries now, I begin to think that I must be a bit of a weak sister when I feel the pinch so quickly! Certainly I give thanks for the training, if not always for the experiences themselves. Yet it would hardly be possible to get one without the other."

Seven people sat down with Martin in the Lodge dining room next day for Christmas dinner. Meeker wrote back to him on December 28: "I am well aware of what you are up against—and how I praise God for the strength of the spirit that has been made manifest through you in meeting all of these problems." A practical man, Meeker included a word of advice regarding Martin's knee. "What are you doing for it?" he asked. "It is possible that a wrapping of cotton, held in place by gauze or sheeting, which will allow overnight soaking with a saturated solution of Epsom salts, might prove helpful."

As the Lodge family grew in numbers, Martin's living room became too small for the meetings; instead they were held in the log building behind the Lodge which Syd Stephenson had built at the turn of the century. After serving as a carpentry and blacksmith shop for many years—and later housing the Lodge generating plant—this impressive structure, with its high-peaked roof and thick, golden logs, found yet another use.

In 1950, wishing to share their experiences with a wider audience, the Lodge community began publishing a small magazine named *Northern Light*. Ernie Taylor, who was to become a long-time resident of the community, wrote in the new publication:

Ever since I first visited the folk here at 100 Mile House many months ago there has been a desire to come here on a more or less permanent basis. At first I could not explain why I had this desire, and yet it remained constantly with me. With each visit, I became more and more sure that this was where I belonged. Now, the question that has occurred to me so often, "Why are we here?" is slowly but surely being answered as each day passes. We are not here for ourselves, to escape from life in the world, or to find peace and contentment for ourselves. If that were the only reason for coming here I am quite sure that our wishes would not be fulfilled, for we cannot accept the privilege of being here without also accepting the responsibilities that go with that privilege. We are here, first and foremost, to learn to serve God, and thereby serve our fellow man. We are here to learn to live with integrity and to love one another, which has never actually been done in the affairs of man, or the world would not be in the state it is, with so much suffering and unhappiness.

In the same issue, Martin outlined the purpose of *Northern Light*: "To point to a new vision of life, in a world so full of darkness and disillusionment; to bring encouragement, to all who would share in this new vision, that there is a solution to every problem, both individual and worldwide; and to inspire confidence that these solutions may be experienced here on earth while we now live." The words summed up his own approach and purpose in life.

15.

A LIGHT IN THE EARTH

Many times over the ages an individual has emerged on the planet and brought inspiration, a light, to others. But all too often that was all that happened. The light-bearer went his way in due course and the candle he had lit soon sputtered out—there was no means whereby the flame might continue to burn.

For seven years, from 1947 until 1954, Martin and Lloyd Meeker worked together to fill the vital need for a protected place on earth where others might have opportunity to awaken to spirit just as they had done. This was the purpose of Sunrise Ranch and the grouping at the Hundred—to provide a setting, or atmosphere, where this could happen. Such awakening is easier, in many ways, when one shares the process with others; as when a team climbs a mountain, there is the advantage of comradeship, mutual encouragement, and shared purpose.

And so while Meeker provided, at Sunrise Ranch, an example of the efficacy of spirit that was utterly compelling to those who were open to it, Martin, at 100 Mile House, did likewise. The bond between them was not based in some well-meaning endeavor to get along together or to respect a common set of principles. It was based in their absolute loyalty to the true nature and character of spirit. Because they held this concern supreme, the two communities could continue to expand in a safe, orderly manner; a protected place for spirit could become increasingly evident.

The need for patience, compassion, and understanding was ever present. Many times those whom Martin was assisting and who had been emerging into a new experience of selfhood would suddenly forget what they were about. They would indulge in an emotional spree of some kind, reacting to some person or situation, and lose their new vision and awareness. There would be no alternative, in such case, but to pick up the pieces and begin again. Although it was difficult not to feel disappointment when this occurred, Martin could never allow those feelings of disappointment to control him. Some people never did emerge from the childish state of wanting to have their own way all the time. But there were others who kept on growing and maturing into a larger, more transcendent perspective; here was reward indeed.

I often noted in Martin that curious blend of gentleness and strength which is part of a truly balanced expression. He never argued or fought with anyone, but when necessary he could be as firm and unmoving as a rock. I recall an instance where a woman in our grouping at 100 Mile House had begun to challenge his authority—surreptitiously, though, not making it too obvious. I am sure Martin was well aware of what was happening behind his back, but thus far he had not been able to get a hand on it. The matter was serious, since the person in question was working hard to spread her feelings of criticism and judgment to others.

Well, finally an opportunity arose to bring something to issue. Martin had begun speaking at one of our meetings about how there is always design, and focus, in spirit. Life doesn't work in a haphazard fashion, according to personal whims or fancies, he brought out: life works the way it works; and just as the human body has a head, a point of focus, so a spiritual body, too, has a head. It wasn't long before an intense current was on the move, to the point that the woman who had been sowing her seeds of dissension and rebellion could stand it no longer. She was a striking woman, with thick black hair and a strong jaw. I can visualize her still as she stood up, eyes flashing, and erupted with a torrent of angry words, criticizing and condemning Martin in the most clear and exhaustive fashion. She also made it clear that she considered herself much more capable of providing spiritual leadership than he was.

The scene was almost reminiscent of a gunfight in the Old West—except that Martin clearly had no intention of getting into a fight with anyone. He remained quiet and still as the woman spoke, not trying to argue or remonstrate, although it was obvious that he was moved by the harsh words flying at him across the room. When she paused, running out of breath, he said simply, "Such attitudes and behavior do not belong in this place. You are free to leave." And that, as far as he was concerned, was the end of the matter. He strode from the room without another glance, leaving a stunned congregation behind him. The person was gone by the following morning. Martin had extended considerable patience, actually, in this situation, but there comes a time when a line is drawn—when the actions of one person begin to adversely affect others.

On an April afternoon in 1950 two dust-stained, heavily laden cars entered the arched gateway leading to the 100 Mile Lodge, one driven by Lloyd Meeker and the other by Martin. They had journeyed many miles together across the huge expanses of Wyoming, Idaho, Washington, and British Columbia, bringing others with them from Sunrise Ranch. John Oshanek, a chiropractor from Winnipeg who had joined the Lodge community the previous year with his wife and family, hurried to get his movie camera to record the event.

Not long after his arrival at 100 Mile House, Meeker brought an unusual opportunity to the attention of the Lodge family. It concerned a meeting to be held in Toronto between a Canadian chiropractor named Albert Ackerley, and another chiropractor, George Shears, from the United States. The meeting promised to be significant, because of the substantial nature of Dr. Shears' influence within the American chiropractic profession.

Shears, a former professional baseball player, was a man of stature, physically and in every other way. In 1939 his own integrity had led him to a recognition that life is a gift from God, and therefore is without price. He acted on his recognition. He stopped charging the many patients who attended his clinic at Huntingburg, Indiana, and instead offered his services on a giving basis, inviting people to contribute what they felt was suitable, or what they could afford. Other chiropractors followed his lead. In 1940 Dr. Shears established GPC Servers, the letters standing for God, Patient, Chiropractor.

In 1948 Shears came to a further realization. He recognized that it was not the physical chiropractic adjustment itself that was of primary importance in the release of life's healing power. With his friend Ackerley and others, he began experimenting with a new kind of adjustment that did not require physical force but rather allowed the current of life moving through his hands to bring about the necessary alignment of the neck and spine. At about this time Dr. Ackerley was introduced to Meeker's literature by John Oshanek and found an immediate agreement with it.

Shears called his new technique "no force" adjusting. In some ways it was similar to the approach which Meeker had developed in his own work with healing, and which he called the "attunement" process. Meeker emphasized that no one can heal another, but that all healing comes from within the body itself through the working of the life forces under the government of spirit.

Through what he termed attunements, Meeker offered specific assistance and encouragement to others so that there might be a greater relaxation and alignment with the currents of life already on the move in all people. "Attunements cannot, as such, make you well," he wrote. "Only God can heal you. Only you can let God's will be done in you. But you need help in learning to let God control your body processes. You need help in learning to live in constant attunement with God's life forces." A typical "attunement" lasts fifteen to twenty minutes. It does not usually include physical touch. The one giving the attunement does, however, use his or her hands to provide a current of healing, into which the other person is invited to relax. In this way the life current moving through both has opportunity to intensify. I have received hundreds of attunements during my own association with the Emissary program, and find them a potent, natural means to connect more deeply with my own source, releasing the blocks that prevent or hinder health.

There were other similarities between the GPC movement and the Emissaries. Most notably, Meeker had also conducted his own program on a giving basis since its inception, making no charge for his literature or for his healing work.

Now there were indications that Dr. Shears was interested in aligning himself and his movement more closely with the Emissaries. As he reviewed the matter with the Lodge family, Meeker underlined the importance of the proposed meeting in Toronto and said that he and Martin were considering flying East so as to offer some input. What did the Unit think of this possibility? Support for the undertaking was unanimous, and the two men left almost immediately for Williams Lake, where they boarded a DC-3 on the first leg of their flight to Ontario.

The meeting with George Shears proved to be a milestone indeed. Later that year, following the annual GPC convention and lyceum at the Palmer College of Chiropractic in Davenport, Iowa, Shears led a group of about forty GPC chiropractors to Sunrise Ranch to share in an intensive five-day seminar. Sitting under a cottonwood tree in the sweet, warm Colorado outdoors that he loved, Meeker told those who had come:

I am not so much concerned about religious considerations as I am concerned about a recognition of the spiritual facts of life. I am not concerned about anyone's opinion, whoever it may be. I am concerned only with Truth, the truths of being. Truth is not something that can be modified on the basis of human opinion or human likes or dislikes. Truth is something that can be discovered. Consequently our approach must be on a scientific basis; we are not interested in fads or fancies. We are not interested in anything that is fanatical. There must be balance; there must be reason; there must be a sound basis for the development of understanding. So it is not religion, as such, with which we are concerned. We are interested in Truth, and we find that Truth is so boundless, so far reaching, so beautiful in all its various phases that we can well believe that it will take eternity for us to explore it adequately . . .

The seminar was a great success and prepared the way for the introduction of regular "art of living" courses at Sunrise. As they expanded and proliferated in North America and elsewhere, these classes in practical spirituality helped people from all over the world develop a more effective and creative approach to life.

During the war years Martin had helped write the constitution for the Cariboo Cattlemen's Association. Now he put his mind to drawing together another constitution. The Emissaries were incorporated as a Society in Canada in December 1951, their purpose ". . . to assist in carrying forward a work of spiritual regeneration of the human race, under the inspiration of the spirit of God."

Toward the end of the same month Martin drove down to Sunrise Ranch with some members of the Lodge family to help launch the first art of living course. It was a day of great fulfillment for Meeker and himself when the six-month class opened on January 2, 1952. Among those who accompanied him on the journey to Sunrise was John Oshanek's son, David, who had been taking part in the men's activities at 100 Mile House. David had experienced the power of attunements in a graphic manner when he fractured his skull in a motorcycle accident at the age of fifteen. Lloyd Meeker was visiting John Oshanek in Winnipeg at the time. The two men went to the hospital where the young man was lying in a coma and shared an attunement with him for twenty-four hours. The accident occurred on a Saturday. Next day, David regained consciousness briefly and went to sleep again. Three days later he was well enough to be brought home in an ambulance, and by the end of the week he had virtually recovered.

Arriving at Sunrise Ranch, Martin found plenty to do physically besides providing an essential point of agreement and stability upon which Meeker could always depend. Help on the ranch was entirely voluntary, and while there was lots of enthusiasm on the part of those who came to live there in those early days, their enthusiasm was not usually accompanied by a great deal of skill. Chiropractors hammered nails. Teachers and artists laid blocks. This lack of expertise showed up painfully in the new meeting hall that was being constructed. The foundation had been laid and the walls built, using cement bricks manufactured on the ranch. But there the project had stopped, because no one knew how to put a roof on the walls without causing them to buckle under the weight.

It was a job for the builder of the 100 Mile Lodge. Martin designed a roof, using a post and beam method that took advantage of the available materials—mainly lumber left over from the building of a government irrigation canal through the valley. There was no money for new lumber. David

Oshanek worked with Martin and saw the project through to completion.

Reminiscent of the quality of his friendship with Lloyd Meeker, another line of agreement emerged between Martin and his son Michael. Because of Edith's ill health, Martin found himself primarily responsible for Michael during his upbringing, a factor which no doubt contributed to the strong bond between them.

Michael had started his schooling in 1940 in the village of Lac La Hache, sixteen miles north of 100 Mile House. From there he went to St. George's school in Vancouver before going to England. While he did not see much of his mother during his later childhood years, he remembers her with much fondness as "a beautiful woman—fine and rather delicate in a way." Not long after he started school in Vancouver, Michael fell sick. He developed a number of childhood diseases all at once and ended up in St. Paul's Hospital. With both his parents in 100 Mile House, he was feeling very much alone until his father came down to Vancouver to see him. "What a ray of light that was to a childish mind," Michael recalls. "He was always very caring and enfolding, though there was firm discipline too. I loved him with a passion."

In the fall of 1946 Michael—just turned eleven—left 100 Mile House with his grandfather to continue his education in England. The parting was painful for both the boy and his father, but their friendship and agreement never wavered during the seven-and-a-half years Michael spent away from home.

During his time in England Michael was directly in the care of his grandparents, Lord and Lady Exeter, and lived with them at Burghley House when he was not at boarding school. They provided a kindly and stable influence for him even though they were of such a different generation. Writing to his mother in August 1946, shortly before Michael's departure for England, Martin touched on some matters relating to his son.

With respect to his prayers, there is nothing cut and dried about this. I feel that the approach to spiritual things should be on an easy and sensible basis, without any tendency to segregate them from the ordinary expression of practical, everyday living. In my view, if a thing isn't practical, it isn't spiritual! Although, of course, there are many spiritual things which may not appear to be practical until they are properly understood. I usually go up in the evening when Michael is in bed and talk things over with him—the events of the day, perhaps. Or sometimes we will read a bit out of the Bible and explore around a little to see what it means. On occasion I may say a spontaneous prayer, or he may do so, or we may use the Lord's prayer, or the 23rd Psalm. There is no routine. I feel that the only worthwhile basis upon which a person can approach the Lord is because he just naturally loves to do so.

Arriving in England, Michael entered Scaitcliffe, an exclusive English preparatory school similar to Lockers Park. It soon became apparent that he was not up to the standard of the rest of the boys. "He is not capable of doing the most elementary questions," his arithmetic teacher complained dolefully in his first report. "At the moment fractions and decimals seem completely beyond him. Well below the standard for his age." Even though willing to work hard and apply himself, there was some question whether Michael would be able to catch up sufficiently to be able to enter Eton in 1949. But to everyone's delight, he did. Lord Exeter wrote to C.R.N. Routh, Michael's future housemaster at Eton College, "I am indeed glad to think that Michael has got one foot on the ladder, and that he has managed to squeeze into Third Form, where his grandfather started life."

Martin wrote to his son every week, maintaining the connection between them and offering a word of reassurance and direction when needed. Once Michael wrote to say he planned to leave the army cadet program at Eton—not one of his favorite activities. The program was designed to prepare students for a two-year period of national service, but Michael had discovered he could quit because he expected to return to Canada after leaving Eton. Martin wrote back in forthright fashion, pointing to the value of following through on something which he had begun. "It wasn't at all what I wanted to hear," Michael recalls, "but the point stayed with me for a long time. It was that kind of thing that made an impression. I knew that whatever came back from my father would really count, even though it wasn't always comfortable."

Michael visited 100 Mile House twice while attending Eton, in 1950 and 1952. In 1950 he had his first exposure to the Lodge community, which had been begun in his absence; he found it fascinating—an eye-opener. His grandfather came with him on that occasion, making his last visit to the ranch he had purchased so many years before. Lord Exeter thoroughly enjoyed his stay and fitted in well with what was happening, although he would not admit to being in agreement with it entirely.

Following his visit to 100 Mile House in 1952, Michael wrote to Sunrise Ranch and asked to be put on the Emissary mailing list. When transcripts of Meeker's talks arrived he would take them to a quiet place at Eton College called Luxmoore's Garden; there he would sit in the sun and marvel at the words and at what they touched deep within him.

Lord and Lady Exeter wanted their grandson to go to Cambridge after Eton. However Michael, with his father, thought it best he return to Canada for a while and see how things worked out. He never did go back to Cambridge. There was so much opportunity and challenge at 100 Mile House that he had no interest in leaving.

Martin met his son in Toronto in January 1954, following Michael's arrival from England, and the two drove back to British Columbia in a new Studebaker station wagon that Martin had bought in the East. Arriving at 100 Mile House, Martin put Michael to work keeping the books at the general store. The business was in poor shape. There was a large backlog of unpaid accounts and the bookkeeping was in disarray. Even though he scarcely knew what a ledger looked like when he started, Michael quickly showed his managerial talent; he became manager within three months, and within a year the store was turning a nice profit.

In March 1954 Edith passed away peacefully in a hospital in Vancouver following an illness. She had brought many beautiful essences into Martin's life since they first met in the Mediterranean, and played an essential part with him, particularly during his early years in the Cariboo. She also gave a precious life to the world in the form of their son Michael.

Four months later Martin saw his friend Lloyd Meeker for the last time also, when he waved good-bye to him and his wife Kathy at Vancouver airport. They were returning to the United States in Meeker's light plane after visiting 100 Mile House. On the morning of August 4, 1954, Meeker, who was a keen pilot and a member of the Colorado Civil Air

Patrol, took off from Oakland airport in his Cessna 172 to return to Sunrise. As he began climbing to cruising altitude over Oakland bay, he encountered haze. The plane stalled, went into a spin, and crashed in shallow water near the shore, taking the lives of all aboard.

Within the short space of five months Martin had lost two of the people closest to him. Moreover, he now faced the full responsibility of continuing the work which Meeker had begun.

16.

A CHALLENGE ACCEPTED

Sometimes in his talks Martin suggested that preparation is always provided to enable a person to deal with the challenges of life when they come. True though the principle may be—and obviously one has to take advantage of opportunities in this regard—when he received word from Sunrise Ranch on August 4, 1954, advising him of the plane crash, it took every ounce of training and backbone Martin had to meet the happening victoriously.

He heard the news from Grace Van Duzen, Lloyd Meeker's secretary. She was in her office at Sunrise when an official at the Oakland airport phoned to tell her of the crash.

"No, that's not possible," was her immediate reaction.

"I'm very sorry, lady. I know this is a shock to you," the man said, and repeated the information.

Besides Lloyd and Kathy Meeker, the tragedy claimed the

lives of Dr. Albert Ackerley, Grace Van Duzen's son Billy, and Sharon Call, daughter of Lillian Call, a member of the Sunrise Ranch community.

Soon after hearing from Sunrise Martin called his communal family together in his home at 100 Mile House and shared the news with them. "There's no point to doing that," he said kindly but firmly to Lena Lucyk as she sat weeping on a sofa. "Everything is the way it is, and we must accept it." This was his attitude throughout the stressful days that followed.

In thinking of this memorable moment in Martin's life I am reminded of a comment Emerson once made about courage: "Every man has his own courage, and is betrayed because he seeks in himself the courage of others." In the drastic situation that now confronted Martin, it was perhaps a kindness that there was no one else he could turn to for help—for courage. The responsibility for doing what needed to be done was totally in his hands; he either accepted that responsibility or he didn't.

As it happened, he accepted it, discovering in the process what is true for each one of us, that he did indeed "have his own courage." Actually, of course, he had been discovering this for a number of years already. He would not have met the hardships of cattle-ranching, or other challenges and tribulations, if he did not possess a good measure of self-reliance—a dogged aversion to quitting. Certainly the difficult circumstances of his earlier life played an important part in preparing him for the present exigency.

What most impressed itself upon him, following the crash, was that the spiritual body which Meeker had brought into initial form should continue to grow and expand and fulfill its mission. This became Martin's chief, and only, concern. Personal feelings of grief and loss had to be set aside. For one thing, his own personal example was all-important. How he acted, how he behaved, would profoundly affect the behavior and reaction of others. But more than that, there was the need to actually assume the mantle of leadership which Meeker had carried hitherto. To his hand, now, was entrusted the direction of that process described in the Emissary constitution as "assisting in the spiritual regeneration of the human race." If Martin was deeply shaken upon hearing of the crash, and he was, it was partly because he recognized the magnitude of the task that now faced him.

Lloyd and Kathy Meeker had three children, Nancy, Lloyd, and Helen. Nancy Rose, the oldest, turned eight just a few weeks before the plane went down. She was with her grandmother, Rosa Groves, when the news came. Rosa was baking biscuits at the Gateway, a property Meeker had purchased two miles down the road from the Ranch. Nancy remembers her grandmother picking up the phone and the look of shock that came over her face; somehow, Nancy knew immediately what had happened. A few minutes later a car arrived and Nancy was taken to her own home, where she found Lillian Call and Grace Van Duzen on hand. "I took comfort from their quiet steadiness in the midst of the atmosphere of shock and the losses they themselves were bearing," Nancy recalls. "Lillian was taking care of details on the phone and it was impressive to me that she remained capable and intelligent in the midst of it all. I knew how deeply she loved my father and mother, and her own daughter Sharon had been on the plane as well. My own heart was traumatized and it took me almost twenty years before I could really release that grief in tears. But there was never a flicker in my mind about the transfer from my father to Martin. It was obvious to me that everything would continue as it should."

Martin conducted the memorial service at Sunrise Ranch on August 8, 1954. Courage firmly in hand, he used the opportunity to great advantage, bringing to bear a dramatically new and different perspective than is customary at such events.

Physical bodies are very important. We could not be gathered here together this afternoon if we did not have them. But the physical form is given meaning, not by reason of itself, but by reason of that which is revealed through it. When life is expressed through a human body, that human body, that human being, becomes meaningful in the world. But the expression of life itself is invisible until the evidence appears in form by reason of the physical body. We are functioning in relationship to this invisible pattern constantly, every day, but seemingly the physical form blinds us to the fact that we have this direct contact with that which is invisible, or has seemed to be invisible.

    The characteristics of life that make a man or woman worthwhile are all invisible until we see them because of some action or word that appears through the physical form. Once that expression of life is absent from the physical form, it has no more meaning, and consequently there is a need, decently and in order, to return the substance to the place from whence it came. That is part of our task this afternoon, but, to my way of thinking, only a very small part, for that which has real meaning is the expression of life itself, and that is no less a reality now than it was before. If we insist that there must be the individual human form to make the expression of life meaningful to us, then we are likely to be very sad and distraught. But when we recognize that that reality, in relationship to each one of these loved ones, is just as real now as it has ever been, or as it ever will be, then everything changes. The sadness that we feel is merely a personal thing to us, in that we enjoyed having our loved ones with us in person. Insofar as the reality of life, of true being, is concerned, they are with us still.

Instead of emphasizing tragedy or loss, therefore, the memorial service brought a sense of comfort and assurance, together with a deep thankfulness for what Lloyd Meeker and the other departed ones had brought during their lives. Many who came from outside the immediate Sunrise Ranch community to offer comfort found that they were the ones who were comforted. They certainly did not find a group of people looking for sympathy, but rather a community of men and women firmly aligned in an attitude of steadfastness and strength. A precedent of victory had been set. But there was a tremendous job yet do.

One person whom life summoned to take larger responsibility was Lillian Call. Born in Milwaukee, Wisconsin, on May 20, 1924, Lillian grew up in Middle America. Her father worked for one of the big steel companies in Milwaukee, starting as a laborer and finishing as a supervisor. All her grandparents were Americans except for one, who emigrated from Sweden as a young man.

Lillian's mother, Marie Johnson, and also her grandmother, had sought the meaning of life for many years. In 1936 they received some of Meeker's literature from a friend and knew immediately that this was what they had been looking for. Although just a young girl, Lillian read the papers too and began writing to Meeker at the age of twelve. When he came to Milwaukee for the first time in 1937 she felt an immediate connection with him.

In the intervening years Lillian married Gregory Lee Call,

but they later separated. In 1948 she visited Sunrise with her grandmother and daughter Sharon, and decided to stay. She had studied music at college and seized the opportunity to organize and direct the first choir on the Ranch; she also took a key role in the women's activities and in the overall development of the community.

It was obvious to Martin that he could not handle his new responsibilities all on his own. He needed someone to stand with him, someone who would be capable of providing a clear agreement and counterpoint in the challenging days ahead. It soon became evident that Lillian Call—with whom he had been friends since 1949—was that one.

They did not come together on the basis of "falling in love," although there was certainly love between them. What drew them was their mutual love for a higher design and purpose—their concern that the work which Lloyd and Kathy Meeker had begun should continue and come to fruition. Given this shared purpose, it was easy for their contrasting personalities to blend. For Martin was a quiet, contained man, with a dry sense of humor, while Lillian is energetic and exuberant. Over the years of their marriage it was a pleasure to observe her love for Martin and her focused care and concern for him. Yet this was never exclusive. Her warmth embraced everything and everyone around her.

In a recent interview Lillian shared some thoughts about the man she married:

Yes, Martin's expression was very different from mine. We laughed about that sometimes. However a certain friendship had developed over the years, until in 1954 it seemed logical that we were the ones to work together and take the responsibility. We were both willing to do that and meet whatever challenges would be ahead of us. And I might add there were a few—not so much from a personal standpoint but in simply rebuilding and continuing on with what Lloyd Meeker had begun.

Martin liked to do things in a straightforward manner. He appreciated the simple beauty of a moment and not all the ramifications that people seem to pour into a situation sometimes, that aren't really needful.

I think he had lots of opportunities to become discouraged in leading a group of people who didn't always see eye to eye. There were many times when he could have let his emotions come to the fore, but I never saw that happen. His heart never ruled, neither did his mind. I always found his beautiful spirit ruled the situation, as we know it is supposed to do.

I appreciated the precise way in which he always did things— driving a car, flying an airplane, doing anything. He always took his time. He reminded me, in that sense, of my old Swedish grandfather, who used to say, "If anything is worth doing, it's worth doing well." Even when he was pounding in a nail my grandfather would take every care to put it in properly, exactly where it belonged—he was a cabinet-maker by trade. And Martin reminded me of him when he did things like that. Precise and careful, an artist at anything he did—even cutting our lawn. You know, for years he cut it certain ways and took care of it. I would see him cutting it in one direction and then the next time he would be going another direction, diagonally, to keep it looking right. It's a beautiful lawn.

Another thing was his simple adherence to the laws we all have to live by—the law of balance, for instance, in the body. In later years, when he played tennis, he recognized his limitations and worked well with his body in relationship to balance on his feet, and so on. He acknowledged the laws of life and lived by them. He never tried to do something silly. He never did when we drove those thousands of miles or flew those hundreds of hours. We never expected him to do something silly, because life has laws and he abided by them. That's why human beings are always in trouble. They are always defying the laws of life.

The two were married at the government agent's office in Kamloops, British Columbia, on September 3, 1954. Conrad O'Brien-ffrench and his wife Rosalie acted as witnesses, and Martin's son, Michael, also attended. Lillian Cecil found herself with an instant family; her husband had become guardian for the three Meeker children, and she was also now Michael's stepmother.

Martin had been a respected rancher and spokesman for the cattle industry in British Columbia for more than twenty years, but much as he loved the ranching life there was no way he could continue to be actively involved in it now that he was directing the Emissary program. As a first step in freeing himself up, he resigned his position as president of the Cariboo Cattlemen's Association, having served in that capacity since 1943.

17.

DEMANDING YEARS

With Lloyd Meeker gone, the care and development of both the 100 Mile Lodge community and Sunrise Ranch fell to Martin. Each member of these two groups was important. From Martin's standpoint each one had a unique part to play in fulfilling the purposes of spirit on earth. To be sure, the Unit members had their quirks and idiosyncrasies, but they had also shown a willingness to let the spirit of love and truth have its way in practical, down-to-earth terms. And that, in the final analysis, was all that mattered. Given that essential openness and the willingness to "keep on keeping on," the refining fire of spirit would do the rest.

For several years, until he learned to fly in 1961, Martin and Lillian drove back and forth between 100 Mile House and Colorado at least twice a year. It was a fifteen-hundred-mile journey each way, taking three or four days. Often they would bring with them one or more of the Meeker children and their own daughter, Marina June, who was born on June 16, 1956. It wasn't unusual for there to be a freshly laundered diaper or two hanging out of a window to dry as they journeyed through the western and northwestern states, stopping for the night in places like Grand Coulee, Washington; Boise, Idaho; or Rock Springs, Wyoming. Though it meant a lot of driving Martin always appreciated the opportunity which the trips provided to share some time with his family away from the demands of the two Units. Also sharing in these journeys was Grace Van Duzen, his personal secretary. She had been Lloyd Meeker's secretary for many years but gave her unwavering support to Martin when the transition occurred.

A sensitive, acute woman with a keen interest in drama and writing, Grace has read widely in her studies of the history of man. Over the years she has offered a unique blend of friendship and encouragement to many people, especially children and young adults. She first met Meeker when he spoke in her hometown of Pittsburgh, Pennsylvania. She had been a student of the Bible for some years before that, searching for the meaning that she knew was there. "The first thing I heard him speak about was Revelation, and it was heaven," she recalls.

But how I met him pointed out to me that you have to be true to your own instinct—no one else can do it for you. I was working in the city. A close friend of mine also worked in town, and we would often meet after work and have dinner together. I didn't know anything about Lloyd Meeker giving a lecture in Pittsburgh. But on the day that the meeting was to be held I had this strong urge to

connect up with my friend. I kept wishing she would call me at the office, and when she didn't, I finally called her. It turned out that she had arranged to have dinner with a friend and then attend this meeting. She didn't want to let me know about it because she had taken me to a number of metaphysical-type meetings and they had always turned out to be a disappointment. But I had a strong urge to join her, and persisted in this—which was not at all like me.
Finally, at dinner, she told me where they were going. I knew at once I had to be there. So I went, and it marked a new and vital beginning in my life.

Grace Van Duzen went to Sunrise Ranch with her son Billy in 1946 for a week's visit, and never left. She met Martin for the first time the following year.

The years following Meeker's death were demanding ones. The emphasis lay on holding steady while people grew and developed spiritually. Many times, as Martin drove back and forth between 100 Mile House and Sunrise, his heart would be in his mouth because he could never be sure what he would find when he reached his destination. The quality of agreement tended to slip while he was away. People who had not yet taken adequate responsibility for themselves might feel assured and trusting of spirit when he was on hand, exemplifying that spirit, but when he left, old doubts and fears tended to rise to the surface.

The two Units went through a natural sorting process. Some individuals welcomed the necessity for growth and change, allowing an increased strength of manhood or womanhood to characterize their lives. Some drifted away, unable or unwilling to accept such a necessity. Still others were willing enough to accept Martin as leader so long as they did not have to put themselves on the spot. He strove constantly to bring home the truth that, as he put it, "We are all called to be leaders."

Anyone who gives any consideration to that which is provided through this program immediately is required to accept a commission of leadership. In our field of leadership we are, properly speaking, pioneers. We are moving through country that has been traversed very little before. If we insist upon staying in the little narrow lanes, the ruts which human beings have worn, we will never lead through that territory where leadership is so needed. If we are accepting our commission to be leaders it is natural to take every opportunity that arises to allow the essential qualities of leadership to be developed, and one of these is versatility. People tend to shy away from anything with which they are not familiar. They say, "This is my field. I can function effectively, or fairly effectively, in this field, but over there I am hopeless." And they usually are! But it is not necessary to be that way. It is simply because the individual has not taken advantage of opportunity. He has kept sheering off whenever the opportunity arose. We cannot be leaders without being versatile.

Martin could utter such words because he was proving them in his own living. He had proved them as a young midshipman on the *Queen Elizabeth* when he delivered the captain's motor car. He had proved them as a rancher, builder, plumber, and carpenter. Now it became necessary to prove them in another area that he would have preferred to avoid.

In taking responsibility to represent the spirit of love and truth on earth he was obliged to articulate that spirit in words, speaking extemporaneously before a wide range of people and in a wide variety of settings, just as Meeker had done. Such expression cannot be predetermined. Some prior thought and preparation is useful, but the words must be spoken fresh and alive according to all the factors that are present in the moment. To do this accurately and consistently requires a great deal of discipline and internal stillness. The ability to "listen," to perceive the subtle pulsations of spirit as they emerge into the realm of thought and feeling, is required. You cannot articulate what you do not hear. As Martin said during one of our interviews, his long, quiet walks with Michael Culme-Seymour at Dartmouth were good training in this kind of "listening," even though he was unaware of that fact at the time. They gave space in which to come away from the busy life of the College and sense the serene, healing influence of spirit.

Martin's speaking delivery was slow and deliberate for a while—painfully so, in the view of some. By no stretch of the imagination was he a good speaker, initially. He became one later, because he persevered. But for the first few years he took his time, because his concern to clothe the spirit accurately was paramount. If the right word was not immediately available to him, he waited until it became available.

"His words came slowly but what he said was always complete and it was always genuine," recalls H.M. "Mac" Duff, then an account executive with a Toronto advertising firm. "I used to admire his guts in sticking with it even though it was difficult for him. Some of those who came to our meetings in Toronto and listened to the tapes of his talks didn't stay long, but there were others who sensed something true and trustworthy and stayed with it. As far as I was concerned, the authority was always there, from the first time I met him."

In August 1954 Martin took over direction of the closing portion of that year's art of living class on Sunrise Ranch. The courses continued to be six months in length until the 1960s, when they began to shorten. Today there are three different levels of courses, each lasting three weeks.

Perhaps reflecting the trauma through which the Emissary

body had passed, class enrollments were small in 1955 and 1956 but upsurged in 1957. Among those attending the 1957 class were Dr. George Shears, the pioneer of the GPC movement, Michael Cecil, Roger de Winton, myself, and the first two students from Nigeria, Timothy Olagunju and Augustine Adesakin. The presence of Timo and Adex, as they were fondly known, capped a series of events that began in the mid-1940s, when a few pieces of Meeker's literature found their way to Nigeria through the hands of a sea captain. In 1946 letters began arriving at Sunrise from that country asking for more guidance and literature. The demand soon snowballed. It became evident that hundreds if not thousands of people in Africa were being deeply touched by the words of a man living in a far-off place called Sunrise Ranch; repeated pleas were made for a representative to be sent to Nigeria from North America.

"Who shall we send? Who *can* we send?" Meeker had puzzled over the question. Now Martin puzzled over it. Resources were small, both in money and in people capable of offering the necessary leadership. Toward the end of 1955 John Oshanek undertook to go to Africa alone. A month after arriving in Nigeria he suffered a stroke and died suddenly in Jericho Hospital, Ibadan. "How are the mighty fallen," the head of the Emissary grouping in Nigeria cabled Martin. "Another superintendent hopeful or hopeless? Please reply urgently." Martin replied: "No superintendent at present. Harvest plenteous, laborers few. Work and pray with us."

It was another setback for a movement still in a tender, embryonic stage. Martin sent a personal message to all on the Emissary mailing list:

When we are faced with such a tragedy as this, the question "Why?" tends to arise. We may easily recognize a number of more or less obvious reasons in the outer sense and, in any case, I do not intend to delve too deeply into the matter, but it is always well that we should pause and search our own hearts. If in fact those who share this program are one body, or are in the process of being received as members of that one body, then obviously we do have great responsibility toward one another. The lives of others are in our hands, even as our lives are in the hands of others. What is the nature of the influence which we exert moment by moment throughout the day? So much depends upon the measure of our faithfulness and integrity.

Both Timo and Adex had worked closely with John Oshanek while he was in Nigeria, Timo serving as his secretary and interpreter. Their presence at the 1957 class at Sunrise—singing, working, and cooking the occasional porcupine on the rimrock as an especial treat for everyone—was a glad and colorful testimony to what Oshanek achieved during his brief time in the country.

Necessarily, Martin devoted more and more time to the needs of the Emissary program. While he continued to take a close, personal interest in all that happened in the village of 100 Mile House, others, such as his son Michael, and Ross Marks, increasingly provided the day-to-day leadership that was needed. 100 Mile House was continuing to expand at a great rate. The boom had hidden advantages, Marks discovered, when an attractive young redhead arrived from Summerland, British Columbia, to teach at the new 100 Mile House elementary school.

Marcia Harvey—as she was named—arrived in September 1952 and stayed at the Lodge before finding a place to board in the village. Marks and his friend David Oshanek—two handsome, eligible bachelors—shared the pleasant task of greeting her and carrying her luggage up to her room when she first arrived. For a few weeks they also shared the privilege of taking her out for sightseeing trips and to the occasional social function. However, it became evident after a while that Marks had the edge in Marcia's affections.

"He was intelligent and fun and always had something to say," she recalls. "There was a sense of purpose there too which was very attractive. I remember that I was surprised to find a person like him in a place like 100 Mile. I hadn't wanted to come here at all, but the school superintendent twisted my arm, telling me how much I would enjoy it. At first I thought Ross was here for the summer, but after a while he explained to me about the Emissaries and I realized the real reason he was here. It didn't make a lot of sense to me at first but he obviously loved it and loved what he was doing. I had a feeling that Martin looked upon him as a son.

"Ross proposed to me while we were walking in Stanley Park in Vancouver during Easter, 1953, and I said 'yes.' Mind you, there were a few times afterwards when I would think to myself, 'Here I am, changing my whole life-style, my thinking, even my name,' and wonder why I was doing this! But we went ahead and got married and everything worked out."

Michael Cecil had continued to manage the 100 Mile House general store since his return to Canada in 1954. Early in 1956 he took on a larger responsibility. As the Kamloops *Daily Sentinel* reported in an enthusiastic front-page spread:

Maintaining its record as the fastest growing settlement in Canada, the 100 Mile House will see the opening on Friday February 3 of the ultra-modern 100 Mile Food Centre, a food supermarket which will bring to the upper country all the comfort, convenience and variety of city shopping.

> *Under the management of Michael Cecil, son of Lord Martin Cecil, the new building has an overall floor area of 4500 square feet, and will require a staff of at least ten ... with the imminent construction of a large hotel, and the recent opening of a hardware store, the fantastic progress of 100 Mile House shows no sign of abatement, and the 100 Mile Food Centre owners look forward confidently to another unprecedented year.*

The continued expansion of the village called for more of the same initiative and pioneering spirit that had been required of its founder. "If it needed to be done, we did it," says Marks, speaking for all those who played a part in the early development of 100 Mile House.

One of the first community projects was the formation of a fire department in 1955, following a fire which burned down the Bridge Creek Estate's garage and shop. Local residents decided that something had to be done. They borrowed some money from the bank and bought a portable pump and some hose; the equipment was put in the back of a pickup truck when the fire alarm went off. A monthly assessment was added to the residential water bills to provide funds.

The following year, an old civil defense fire truck was located near Seattle and brought across the border. Marks, who served as fire chief for the first ten years of the department's life, recalls the first time the brigade went into action: "It was total confusion. None of the men had had any training and they didn't have a clue what they were doing. In one instance people headed into the fire holding on to two ends of the same hose. They were hooked up to each other instead of to the water!"

At a dinner given in his honor by the 100 Mile House and District Historical Society in later years, Martin spoke of the importance of the pioneering spirit. Referring to the early years in 100 Mile House he said:

As you know, there were no so-called modern conveniences at that time. There was no electric power, no running water, none of the civilized assistances that are taken for granted in these days. One had to depend upon oneself. It was necessary to develop some adequate character in order to handle the situation, without any expectation that someone else would take care of you. Nowadays it seems that we are all taken care of, from the cradle to the grave, so to speak; we don't have to bother. I think we have lost something in this—maybe we sold our birthright for a mess of pottage—because there is a need for this sort of character, I'm sure. All those who were pioneers in the old days were very much aware of the self-reliance that was required. There were neighbors, of course, helpful neighbors, but they were few and far between. One had to handle things oneself.

And so, as I say, I believe we have lost something over the years in this part of the world, even though we seem to have gained something. It is so much more comfortable than it used to be. We can, with a touch of the finger, turn on the electric light. We have everything that one could wish for in this beautiful civilized world we have. But let us preserve what might well be called the pioneer spirit, which doesn't expect to be provided with everything by other people, particularly I suppose these days by the government.

That is the direction which leads to abject slavery, and we are well on the way. We apparently are inclined to move in that direction voluntarily, because there is always the demand for something more, something more to be given us. I think it is more important that we should look around and see what it is we might give to others, to our community, to whatever is necessary in the field of our particular responsibility . . . this has been my concern in the development of this community. I could have held everyone up for ransom, I suppose, but I never did that. This was proven out in many cases where we sold property and very shortly thereafter it was sold for considerably more. So we have a beautiful part of the world to preserve on the basis of our own integrity and willingness to take responsibility for ourselves. I am sure that all of you have done that over your lives. Let us not see it cast aside.

18.

A MAGNIFICENT WHOLE

As greater stability and trustworthiness developed at 100 Mile House and Sunrise Ranch Martin was able to venture beyond those two core communities, carrying the invitation of spirit directly to a troubled, confused world.

What was the invitation he brought—that Emissaries everywhere continue to extend? I can only speak from my own experience. I listened to Martin's addresses at 100 Mile House for thirty years. For several of those years he spoke on Wednesdays and Saturdays as well as twice on Sundays. In 1963 he passed the responsibility for the midweek presentations onto the rest of us, but continued to give the Sunday ones.

I know that it was not habit, or a sense of duty, that compelled me to sit with him for all those hours. He was calling something to my remembrance that was infinitely attractive and satisfying: a remembrance of my own true stature, my own place in a magnificent whole forever pulsing with serenity and beauty. In every word he spoke, and every talk he gave, he invited me to remember that I was more than just "Chris Foster." That name describes my outer, physical self, with its various hereditary traits and acquired characteristics, but it does not describe the spiritual being who is incarnate in this physical form.

The unfortunate effect of most religions has been to convince people that God—if such an entity exists—is somewhere distant and far-off. During those hours which we shared together, Martin showed me that the spirit of God is not somewhere else. It is exactly where I am. That spirit is me. I am that one. I am not a finite human condemned to live a few brief years on earth and then die. Yes, the outer body known as Chris Foster will die—but then Chris Foster has been metamorphosing for quite a while already, as the one who I truly am emerges increasingly into expression.

I once portrayed these things in a novel that I entitled *Bearers of the Sun*. Written with Martin's interested encouragement, the book told of four "Sun Beings" who incarnated on earth in 1920 to carry out a commission of great importance to the planet. One incarnated in Russia, one in England, and two in North America. At first I merely planned to show my four characters awakening, step by step, to their true identity as Sun Beings. It was Martin who suggested that if they were friends in the Sun, as I had postulated, then logically they would find themselves drawn together to become friends on earth. And so that was how I shaped the book. I worked things out so that through a series of seeming coincidences, the four finally met one another—

at a cocktail party in New York—becoming aware of a friendship that transcended national and cultural borders.

This is a metaphor for a process that is unfolding on the planet now. More and more people are realizing their identity in spirit—their membership in the quiet, melodious order of the universe. Realizing this, all the endless human striving to achieve, to succeed, to become a better person even, begins to be seen as unnecessary. The one who is incarnate is already perfect. True, there is a need to train and refine the capacities of mind, heart, and body which make it possible for the incarnate one to act and express on earth. But the incarnate one himself or herself is already perfect. "Be ye therefore perfect, even as your Father which is in heaven is perfect," Jesus said—all that is necessary is to let that perfection find release. Martin called this spiritual expression. "Never underestimate the power of spiritual expression," he would say. The power of such release radiating from even one individual is immense: allowed to radiate through a cohesive body of people, it becomes incalculable.

Those who see only the troubles and catastrophes facing humanity look askance at such a simple approach as this. They think the only answer is to battle the problems that are so clearly visible all around—to develop yet more clever and ingenious strategies of the kind that produced those problems in the first place. Such people often accused Martin of being indifferent to worldly troubles. He was not indifferent by any means. He just knew how life works—where the cause of human troubles lies, and where the solution lies. He once wrote a remarkable letter in which he addressed this apparent dilemma:

Clearly, from the usual compassionate standpoint there is a need for help in the world. Great numbers of people, including children, are suffering and dying. Such tragedies have been the lot of humanity all down through the ages.

In every generation there have been compassionate people who have done many heroic things (usually unsung) to alleviate the hell which human greed and fear and lust for power have produced on earth. There are indeed such angels of mercy today.

In spite of all that has been poured out to comfort and to heal, the need for comfort and healing continues to multiply in the world. No matter how humane the effort may seem and no matter how many may be blest in whatever measure, there are always vastly more who find themselves in a desolation without hope. Looking at the situation honestly, one would have to acknowledge the ultimate futility of the ways of human compassion. But there is another way. Very few on the face of the earth have had the direct opportunity of touching this other way.

There have been a few who have now touched it, you amongst them. The invitation that is extended to these few is vastly greater than anything that could be achieved by sincere human beings who are attempting to stem the destructive flood that is engulfing the entire earth.

If a person has not had the opportunity to touch this one way in the direct sense, then if they have integrity they must necessarily do what they are able to do to the highest of their understanding and their compassionate view. This might compel a person to go to somewhere like Cambodia to offer help. However, such as do this almost inevitably must be consumed by the hopeless futility of it all. That they go ahead and do what they have to do in spite of it all is heroic. However, nothing thereby would have been done to change the course of events which the very state of humankind makes inevitable.

There is only one true way. And those who have touched the manifest evidence of it in a direct sense have the greatest possible responsibility placed in their hands. This one way requires that the individual reveal the truth in his or her living now. If there have been relatively few who have accepted this invitation, this does not mean that the invitation is not genuine and the way is not open for the power of spirit to transform the world.

In the fall of 1960 Martin drove out of the driveway of his home at Sunrise Ranch to begin a four-week visit to the Midwest, eastern Canada, and the eastern United States. Accompanying him in his green, befinned Chrysler were his wife Lillian and Roger de Winton, a former Royal Canadian Navy officer and bank manager who made his home at Sunrise after attending the Emissary class there in 1957. The 5,270-mile journey was significant because it set the stage for a great deal of traveling and public speaking in the years to follow; from this point on, Martin and Lillian would tour different parts of the North American continent regularly year after year.

Roger de Winton told the Sunrise Ranch family after their return, "How can I convey in words the excitement of this trip? It was the most electrifying thing you could imagine. You realized that sometimes people did not understand what Martin was talking about, some of them, but all the way along the line there was this tremendous, vibrant response." Lillian Cecil commented, "An example of the high level of interest that we found throughout was the public meeting in Toronto. People came from various backgrounds and with varying degrees of understanding. Some were hearing Martin for the first time, yet there was the most wonderful warm current of interest, an awareness that here was the answer, here was the end of their searching."

Some two-hundred-and-forty people attended the meeting in the King Edward Sheraton in Toronto, including many chiropractors. These came along at various stages of the proceedings as soon as they were able to close their clinics; one showed up at ten o'clock when there was no one left but the janitor. After various meetings and talks in Ontario, and also in Montreal, Martin and party headed south for the U.S. border. It was a hot fall day as they drove through Quebec. They pulled up at a rural gas station and asked the elderly attendant if he had anything cold to drink. He beckoned to Martin and de Winton to follow him, and they disappeared into a back room where the man drew two glasses of homemade cider from a cask. It was delicious, cool—and highly potent; how potent, they did not realize until they got back into the car and continued on their way. They drove down to the border very cautiously indeed.

They met many members of the cooperative chiropractic movement during the trip, especially in New England. In New Hampshire Martin spoke to a group of thirty-five people in a chiropractic clinic in North Conway. While he did not pay much attention to it at the time, the room was small and most of those present were smoking. Later that night he found he had completely lost his voice because of that smoky atmosphere—this, with some important meetings immediately at hand. "You may have to take my place," he whispered to de Winton, whose pulse quickened a notch, for he had little experience of public speaking at that point. However, with some rest and attunements Martin's throat cleared. Next day they drove down to the southern part of New Hampshire and visited a bustling chiropractic clinic in Derry operated by Dr. William H. Bahan and his two brothers, Wally and Paul.

Bill Bahan, a native of Massachusetts, was brought up in a family of five boys and four girls. Once his mother bought him a new coat. When he came home on the very first day without it she asked, "Where's your jacket?"

"I gave it to another boy," he replied, as if it was the most natural thing in the world.

A nephew of Dr. George Shears, Bill Bahan decided at an early age to be a chiropractor like his Uncle George, and graduated from the Palmer College of Chiropractic in Davenport, Iowa, in 1949. He was a buoyant, unique man. Once, when a woman was reciting a long list of all the things that she thought were wrong with her, he stopped her in her tracks by asking, "How's your knee?" Her knee was fine, which was why Bahan drew her attention to it: he wanted to emphasize that there is always something right in any situation—and in the field of health, as in most other areas, that is the best place to start. Given his uncle's deep love for Lloyd Meeker and Martin and what they represented on the planet, it was not surprising that Bahan followed suit in that regard too. Soon after meeting Martin at a family gathering in Methuen, Massachusetts, he began to play an important role in the development of the Emissary program.

From New Hampshire, Martin and his companions drove to Massachusetts and New York, concluding their tour with a visit to Huntingburg, Indiana, Dr. Shears' hometown.

Such a trip is made up of countless little details. Roger de Winton remembers the clever way in which Martin packed, arranging the luggage in the trunk in such a manner that when he arrived at a way-point, it was only necessary to remove a small amount of baggage from the top; everything else was left untouched so that they could leave quickly the following morning. Then there was the time they were visiting a family in London, Ontario, and Lillian sent for some yogurt for a small boy who had a tummy ache; it was just what the child needed, and he soon got well.

Martin impressed de Winton with his intuitive ability, a kind of sixth sense. The latter recalls: "Sometimes, when I was taking a turn at driving, he would seem to be asleep, but then he would wake up and ask me how fast I was going. I would be going over the limit and he would say, 'I think you'd better slow down.' The country would be flat as a pancake, with nothing in sight, but sure enough, in a few minutes a police car would appear out of nowhere." I can relate to such a comment. Many years ago, when a group of us were on a picnic at a lake near 100 Mile House, I lost my glasses. Realizing that they must have fallen out of my pocket as we walked through some dense clumps of bushes, I despaired of ever finding them. Martin went on a quest of his own, however. He walked back through the bushes and calmly plucked the glasses out of some foliage.

Martin and Lillian returned to 100 Mile House in November 1960. Not long afterwards John F. Kennedy was inaugurated as the thirty-fifth president of the United States, the youngest person to be elected to that office. His words, spoken with a crisp Boston accent from the cold, windswept platform at the Capitol, touched the hearts of millions.

Martin was always concerned to offer support and agreement to anyone who revealed evidences of true character and purpose. In his Sunday morning address on January 22, 1961, the day following the inauguration, Martin read the text of the President's speech, with its strong call to confront the problems facing humankind, and added some strong words of his own:

... In these words we can find a wonderful spirit of agreement. It may be said that they are just words. We recognize tyranny, poverty, disease, and war as evidences of man's failure to be true to the larger design of life. We recognize that these things cannot be eliminated by struggling with them, but we do recognize that they must be eliminated. It is through the working of God's power in the achievement of God's purposes that such things are caused to pass away: the tyranny of self-centered human beings, human beings determined to get their own way regardless of the larger whole. Poverty? Poverty is a man-made thing. Perhaps we would be particularly aware of the poverty of spirit in man, the poverty which is so universal insofar as man's awareness and acceptance of the spirit of God is concerned. Under the dominion of that spirit there is no such thing as the material poverty which human beings experience so widely in the world today, for all of the riches which God created in this His earth are freely available when they are used rightly—available to all. Disease?—the evidence of the separation of man and spirit: diseases of body, mind, and heart. How shall these diseases be caused to vanish away? Only as man is restored to his place in the whole. No amount of struggling with them will eliminate them. So we surely subscribe to this same intent: These things, these evil things shall be known no more on earth forever.

One spring afternoon in 1961 Ross Marks approached Martin with an interesting thought. A pilot training program was being arranged through the local Flying Club and the Aero Club of B.C. How about he and Martin learning to fly? Martin liked the idea. The cost was reasonable—just $500— and in due course they might be able to buy an aircraft, which would eliminate the long, tedious automobile journeys to and from Sunrise and elsewhere.

Marks recalls his own first lesson in the 100 Mile House Flying Club's new Piper Colt aircraft: "With a certain amount of trepidation, I strapped myself in the 'driver's seat,' peering at the cluster of unfamiliar dials and gauges spread before me. The only things I recognized were the gas gauges and one or two other automotive-type dials. The wheel did look something like a steering wheel, but I suspected that it didn't do quite the same things as its counterpart in a car. I have a vague recollection of bumping down the runway and soaring into the air, all the while attempting to catch the comments of the instructor. Back on the control column—up we went. Forward on the column—and down, down, like an express elevator, with my stomach feeling as if it was in my throat!"

Martin found the solo training period an unusual and welcome opportunity to be alone. "It was something that I had never experienced before, in the same way at least," he once recalled. "I suppose there are times on the ground when something like that might happen, but it's different when you are in the air—you are on your own and in a sense above everything, without any distractions." The two men practised take-offs and landings at an old RCAF airport at Dog Creek, by the mighty Fraser River. One of the runways faced directly over the Fraser, so that a pilot becoming airborne immediately found him- or herself flying at 1,000 feet.

Later in the year, while Martin was conducting a class at Sunrise, an opportunity arose to buy a four-seater Cessna 182. Gordon Maitland, of 100 Mile House, a former World War Two pilot, helped Marks fly the plane down to Sunrise Ranch so that Martin could look at it prior to its purchase. The plane—identified by the letters CF-JYE—arrived at Loveland municipal airport to a warm welcome. A short while later Martin flew CF-JYE to Farmington, New Mexico, partly to get the feel of the plane and partly to buy a second-hand diesel generating plant for the Ranch. Martin's family company, Bridge Creek Estate, bought the aircraft soon afterwards. When the first issue of *Newslight,* the Emissaries' newspaper, came off the Sunrise Ranch press in November 1961, JYE and Martin were pictured on the front page underneath the headline, "Clearing the Air!"

Now, instead of driving to distant places, he and Lillian began to do much of their travel by air—at an average speed, in JYE, of 150 miles per hour. In the 1963 edition of *Northern Light,* Lillian described some of her experiences sitting in the navigator's seat:

March 1962—100 Mile House to Penticton:
My, this is smooth. Now this is the kind of flying I like. I wonder why I was worrying about this trip—I feel like singing.

Penticton to Lethbridge:
Getting a little bumpy—oh well, I feel good today. I won't be subject to my stomach.

Hmmm—it seems to be bumpier—maybe my imagination—hope my stomach remembers I'm not subject to it! Easy there, JYE. How does Martin do it? He looks so calm, and IS, too!

Look at those peaks—the Rocky Mountains—beginning to look cloudy. Yes, Martin? We're changing course. Why can't I find that railroad? Dear God, where is it? Oh, Martin sees it. He certainly has wonderful vision—I could say that again!

Easy, stomach—my, look at the snow so close. Yes, must keep looking at the map and out the window, checking. Better not close my eyes for a moment. We'd better know where we are every second.

Easy, easy. Oh, the map shows only one more peak to pass by—should be clear beyond. Oh, it isn't—a snowstorm. What a turn!

New direction, on to Lethbridge. Oh well, stomach, nothing left now!

On to Sunrise. This is more like it—guess I'll doze while Grace navigates. What's that, Marina? you want to play, don't want to sleep?

August 1962 -- Pendleton to Ogden:
It's rather bumpy, not bad though; rather nice, sitting back here with Marina. I'll doze a bit. What's that, Marina? You NEED an ice cream cone! And where will I get one? We're 9,500 feet up in the air.

Martin became seriously ill at the beginning of January 1963, with complications stemming from his appendectomy in the 1930s. The trouble was not only physical, however. He had been over-extending himself for some time—partly because of the unwillingness of others to take adequate responsibility for themselves—and finally it all caught up with him. "Because this is a body of many members the responsibility can on that basis be handled quite easily," he remarked after the crisis was over, "but beyond a certain point it is too much for one physical body to carry."

Martin's near-fatal illness inspired considerable change and maturing throughout the entire Emissary program. Providing firm leadership in this was his son Michael, who with strong agreement from Bill Bahan, Jim Wellemeyer, Roger de Winton, and others, began to play an increasingly prominent role in coordinating Emissary activities. Indicative of the new thrust which was underway, Michael took the reins when a three-day British Columbia Servers' Conference, as it was called, was held at 100 Mile House toward the end of January. Martin spoke at the last session of the conference, pointing out, with some humor, that he was not retiring, but that he had other work to do beyond caring for those who should be capable of caring for themselves. Swami Premananda, a saffron-robed guru from India, added a colorful touch to the gathering. He was a stocky, crew-cut young man of thirty-two who had given up a comfortable life as a history professor in India to spread his joyous philosophy of love throughout the world. "O Truth, I love Thee," he used to say. "O Love, I am true to Thee." Premananda had met the Emissary movement through Richard Thompson, and had made his first visit to 100 Mile House a year or so previously, finding an immediate affinity with Martin.

The seed of love and respect present between the two men sprouted later in 1963 when Premananda invited Martin to participate with him in convening two public events, at Atlanta, Georgia, and Buffalo, New York, entitled "Awake O' Man." Typical of Premananda's keen sense of fun, he opened the Buffalo gathering by saying it would be "an unconventional convention, for all man-made traditions are untraditional to man's true divine nature." He then introduced Martin, saying he had been "the great source of inspiration to me in providing this opportunity to you all." Presiding over the first session, Martin suggested that those present had been drawn by an inner force strong enough to overcome any problems or difficulties that might have stood in the way. This same compulsion, he said, would continue to offer what was necessary in their individual living of life in the hours and days ahead.

The gathering was unconventional in several ways. While there was a program, the organizers did not follow it closely, and sometimes the titles of the talks were changed at the last minute. At one point Martin got up to speak, then turned to Premananda and asked, "What am I supposed to talk about?"

It was a humorous moment, but it made a point too—that his talks really were spontaneous and in the flow of the spirit.

Reporting back to the family at Sunrise shortly after the Buffalo gathering Martin touched on one of the main themes:

The matter of human egos came up quite frequently in the convention, and this is what causes all the trouble, isn't it? Nothing else. There is not any other problem, actually. Just this, the human ego. It wants to be out there in front. It wants to have its own way. It wants to do as it pleases. It wants to fit in if it finds it convenient. But really to be willing to evaporate, that would be too much. And yet as long as the human ego remains the individual is worthless, absolutely worthless. Spirit is to act on earth; spirit needs a body, a mind, a heart to act through. If the human ego is occupying that instrument, of what value is the instrument to spirit? The ego must be relinquished. There must be a willingness to let it go. That includes your precious pride and your precious feelings. All our human problems go when the ego goes. You cannot get rid of them and keep your ego. It is either one thing or the other.

Swami Premananda died in an automobile accident in India a little over a year later, but his strong agreement with Martin prepared the way for an increasingly close connection with India in years to come.

Bill Bahan, who had already inspired many people to awaken to a greater sense of true purpose, arrived at Sunrise Ranch in April 1963 to attend an art of living course himself; not surprisingly, he brought with him a caravan of six cars carrying eighteen adults and thirteen children, causing much excitement and activity as they were all welcomed and shown to their rooms. At that time Martin was still conducting the courses himself. The following year Bahan began to take over that task.

A further sign of maturing and growth within the Emissary ranks was the purchase of a sixty-seven-acre farm in Epping, New Hampshire, in the fall of 1963. Under the guiding hand of Bill Bahan's brother, Walter, this became the first Emissary community to be established beyond Sunrise Ranch and the 100 Mile Unit. The property was named Green Pastures because of its beautiful rolling fields.

The 100 Mile Lodge had served travelers on the Cariboo Highway for more than thirty years—and served them well. By the 1960s, however, it was becoming increasingly inadequate, from the standpoint both of size and also the type of accommodation it offered. In the fall of 1964 Martin raised the possibility of building a new motel complex, with first class dining facilities and a banquet room, that would front the Cariboo Highway more or less on the site of the old stopping house.

The proposed new development called for some stretching by the 100 Mile House Emissary family. Hitherto they had been tucked away to the west of the main road. Now they would be in the mainstream; moreover, with 46 units, the new development would be substantially larger than the Lodge. At a gathering one evening Martin spelt out the options: "The idea of the motel has reached a point where something rather specific is available for consideration, and it is a question of either going ahead or not. Now, with respect to our movement as a whole, emphasis has recently been placed upon the fact that things are on the move. Even if we wanted to, we are not going to be able to settle down and keep things the way they are. We either keep moving—I am thinking of movement in the sense of what is unfolding from a spiritual standpoint—or we cease to be associated with what is unfolding." The necessary willingness and initiative were forthcoming. In the spring of 1965 local contractor Ian Galpin fired up his bulldozer and began levelling the site for a new motel complex to be named Red Coach Inn.

The same year also saw the incorporation of 100 Mile House. Interviewed by the local newspaper, Martin made it clear that he would agree to incorporation if a majority of leaseholders desired it. "We have a genuine interest in the village, but if the residents feel they would be better off under incorporation then they should go ahead," he said. Reflecting his own attitude to life, he said he thought it would be a mistake to opt for incorporation simply to obtain a per capita grant or because one was dissatisfied with the way things were. "There is no magic formula," he declared. "But if it is the other way—if people want to do a job—then more power to them. If there is a favorable vote the Bridge Creek Estate will support it 100 percent." He promised that if this happened he would sell his land at current market prices to those who wished to purchase.

Once before, when an unofficial vote was taken, incorporation had been turned down. In June 1965 it was approved by a narrow margin; the vote was 61.3 percent, with a minimum of 60 percent being required to pass. A.V. MacMillan, incorporation committee chairman, said the vote showed 100 Mile House had "come of age."

Early in July a large crowd filed into the 100 Mile House community hall to elect an interim village council. Ross Marks and Chamber of Commerce President R.J. "Spud" Speers were nominated for the position of chairman. Marks at first declined, but finally allowed his name to stand at the insistence of his nominator, lumberman David Ainsworth, plus spontaneous applause from those present. The final vote was Marks 51, Speers 7. When a regular municipal election was held later, all the members of the interim council were returned, and Marks was elected the first Mayor of 100 Mile House by acclamation.

19.

A DECADE OF CHANGE

That most turbulent of times, the 1960s, was also a time when the body of human beings which Martin was drawing together as a vehicle for the action of spirit began to flesh out and take on more recognizable form and shape.

Along with the turbulence, the 60s are often viewed as an era of creative and moral upsurge. Joan Baez, a symbol of those years, put it this way: "People were willing to take a little more risk." It was a time of both protest and dreaming, as typified by some of the slogans that made their way around the world—"Make love, not war" or, "Flower power." Betty Friedan published *The Feminine Mystique,* helping to launch the feminist movement. Martin Luther King Jr. led 250,000 civil rights workers in their famous march on Washington. The Beatles were honored by the

Queen, to the chagrin of some members of the British Establishment. The Baby Boomers, of course, fostered the ferment—seventy-six million of them were born in the United States alone between 1946 to 1964—challenging the whole concept of the Vietnam war and the hypocrisy and greed which they perceived in the Establishment. But back of all the protest, the divisiveness, and the dreaming, was another force—life itself, the power of spirit, pushing its way through the rigid structures and determinations to which human beings cling with such tenacity, but which distort the clear flow of life in their own experience. Here is the "hidden player"—the fundamental cause of all change and transformation occurring on earth; though acknowledged by rather few of our planet's inhabitants, life will have its way in the end for all that.

As this movement of life continued, two great forces intensified in the experience of people everywhere: on one hand, chaos and disintegration; on the other, unity and integration.

From Martin's standpoint, both forces were valuable; both, indeed, were essential, working together to make possible the means whereby old forms that had outlived their usefulness passed away, and new forms more suitable to the immediate clothing of spirit appeared.

More than that, he knew—as he criss-crossed North America, speaking the word of spirit in every corner of that great continent—that all was well. All was contained within the working of life's eternal, immutable laws.

Where a person yearned for a greater experience of meaning, of unity, of integrity, that was the direction in which he or she moved. Events conspired to make it so. Where a person was preoccupied with selfish concerns and motivations, ignoring the compulsion of spirit—the compulsion of the universe—then chaos and disintegration naturally ensued. How could it be otherwise?

As the ferment of the 60s continued, Martin and his associates found an increasing openness across the land to the message that they brought. From the firm base now provided by the communities at Sunrise Ranch and 100 Mile House, joined in 1963 by Green Pastures, the Emissary presence reached out and covered the whole North American continent. Art of living courses began to proliferate and increase in their enrollments. Groups sprang up in universities. New centers opened in many cities.

In the East and South, Bill Bahan played an active part in the expansion, as this tongue-in-cheek report from *Newslight* portrayed:

In 1864 the South was invaded by General William Tecumseh Sherman, who led his Union army on a destructive march to the sea. "War is hell," said Sherman, and the South agreed with him. Now, 103 years later, an invitation to unity is being offered on an entirely different basis than Sherman's. "Life is wonderful," declares Bill Bahan as he travels the South from Louisiana to Florida, from the Georgia mountains to the Cajun swamplands, and the South agrees with him. Complete details of Bill's campaign are not available at this time, but scattered reports in from the area indicate that he will no doubt be devoting an increasing amount of his time there.

Many baby boomers found themselves in a dilemma as the decade progressed. If they reverted to the old, familiar values that had satisfied their forebears, feelings of frustration and despair tended to arise—understandably so. Yet the awareness dawned for some that they were hitting their heads against a ceiling—that ultimately drugs and protest led nowhere. Here was fertile soil. For these, Martin's word of truth, balance, and common sense opened a door to the larger experience of identity for which they longed and which inherently they knew did exist.

David Thatcher was typical of many of those who passed through that door and began to flesh out the Emissary program during this period. Born in London, Ontario, he first sensed that he was a "stranger in a strange land" when he was five. During the mid-1960s he went to a variety of churches, none of which attracted him; after reading Jacob Needleman's book, *The New Religions,* he began connecting with various alternative communities. Thatcher was an honors student at college, with a bright future, materially speaking. But he had already plotted all that out and knew that it was not sufficient. His experiments with LSD strengthened his conviction that there was a larger dimension to life, but he knew at a certain point that it was time to quit drugs. Shortly after doing so he encountered the Emissary program in Ontario. Upon visiting 100 Mile House and meeting Martin and Michael he knew immediately that "this was it—this was my family."

If Joan Baez, the Beatles, and Betty Friedan were at one end of a spectrum, Sir Winston Churchill, symbol of the British spirit and lifelong defender of the British Empire, was at the other.

For people like me, who had lived in London as a boy during World War Two, and for a great many others around the world, Churchill's death on January 23, 1965, at the age of ninety, was a moving and poignant moment carrying as much impact as any of the other changes of the 1960s. Here was a man who in England's darkest hour lit a flame of courage that united the will of a nation.

In an address on the morning of January 24 Martin spoke of that example and of its relevance to the present day. Having recounted how England, at the beginning of the Second World War, faced a much stronger and more powerful enemy, he continued:

... And yet this man, Sir Winston Churchill, with the remarkable expression of spirit which characterized him, provided something which ensured ultimate victory. It wasn't a material thing. The essential armaments, for instance, couldn't be conjured out of thin air: they simply weren't there. But something was offered which was, in one sense, quite intangible, the spirit of victory we might say—this is the spirit of love, isn't it? And there was a response to that. Here was a remarkable portrayal of the practical nature of the spirit. It gets the job done. At that time the so-called enemy had the material, the arms, the men, and seemed to be in a victorious position. But that is not the way it proved out, because of the presence and the activation of spirit.

With examples of this nature, it is surprising, is it not, how human beings still refuse to acknowledge the place and the true meaning of this intangible thing called spirit. When what has been called a miracle of this nature occurs, people are impressed, but not to the extent of recognizing how it happened, apparently, and everyone feels much safer if they have their stockpiles of atomic bombs and all the rest laid conveniently away. "These are much more trustworthy than spirit," they say in effect. But it isn't so. Wherever the spirit of love comes to focus and is allowed to find release, so that it is multiplied by activation through others, there is victory, there is strength, there is that which makes possible the achievement of miracles, so called.

A quarter-century later, as all humankind faces what could well seem an even more perilous future, I ponder Martin's words. Is the power of spirit any less present today than it was in the British people in 1940? Is the possibility of a victorious, transcendent experience, individually or collectively, any less present? Surely not. It all hinges upon the willingness to align ourselves with the power of love, with the indomitable power of spirit, as Churchill did, so that the same principles that worked for him and for Britain in 1940 may work today. Our true nature is indomitable. When we honor that true nature in ourselves, no matter how impossible our circumstances may seem, the power of love will triumph. Martin described the process in a prose poem drawn from the clear, azure depths of his own living. He entitled the poem "Thus It Is":

>From age to age
Love's word rings forth,
"The truth is true and all is well,
Unconquerable life prevails."
Oh, man, whose strident dreams
Lead gravewards,
Return to calm and noble
Character of life.
Blaze forth pure virtue;
Depart false ambition's restless schemes.
Busy thought and troubled feeling
Trespass not in virtue's wise serenity
Where firm control and awful power
Eternally abide.
Here earth's pains are healed
And cruel chaos of mind's spawning
Is called again to order and to beauty.

On a brilliant fall afternoon in 1966 Michael Cecil and Nancy Meeker sat down with Martin and Lillian to tell of the love which had grown between them, nurturing in their hearts and whispering of the possibility of marriage. It was a tender moment in Martin's living room as the young couple spoke of their feelings for one another. There was an atmosphere of thankfulness and delight present, an obvious welcoming of what was happening, but no one felt any need to make a decision. Life was bringing something new into form and could be trusted to continue that process. On January 2, 1967, Martin announced the engagement of his only son, Michael, and Lloyd Meeker's eldest daughter, Nancy Rose.

Nancy was nine when she first got to know Michael. With her brother Lloyd she spent the summer of 1955 at 100 Mile House. At that time Michael was busy managing the store. "Being eleven years older than I am, he was at a very different phase of his life," she recalls.

He was very grown-up and already a leader, and because he was so busy he seemed somewhat remote. In a way he was like a brother because he was Martin's son, but in another way he was so much older that I didn't really think of him as a brother at all; he was just himself. And then I remember a phase when I was fifteen or sixteen, and living here at 100 Mile House. At night it was the children's tradition to go around and kiss everyone goodnight before going up to bed. There came a point where I was a bit shy about kissing Michael goodnight. I remember him being very sweet with that shyness, certainly not expecting anything from me, but just giving me a very sweet smile. And then when I went to university—I was only there one year—he would always come and visit me when he was in Vancouver and spend a little time. His presence was always very comforting and reassuring. He carried such a strong polarity in the truth—there was just no question about what was important or what needed to be done. I suppose that during that time I gradually came closer to him in that way, just accepting his polarity, so that when I moved back up to 100 Mile House it seemed very natural for something to open up between us on a more personal basis. It certainly wasn't sudden, and we laughed with each other at the time of the engagement, agreeing that at least we knew each other very well. There were no surprises in that sense. It was already a very close and comfortable relationship which could be the foundation for a new dimension to open up.

As Nancy suggests, Michael's primary characteristic, certainly as long as I have known him, has always been his total concern for the fulfillment of life's purposes. This concern is unwavering, demonstrated in such things as his scrupulous use of time. "Time is of the essence," he once remarked in a conversation many years ago; yet while he forbids himself to waste —be it in relationship to time or anything else—paradoxically he expresses a sense of fun and ease that is thoroughly contagious. It is hard to be too serious around Michael, even though, I may add, he is able to inspire and hold safe the most tender and intimate interchange. I suppose this all says something about the balance inherent in the true nature of spirit.

I am sure others would agree with me that by the time his engagement was announced—indeed, well before that—Michael had earned the respect and love of all those closely associated with the Emissary program. It wasn't based in who Michael was in an outer sense. It was based in his own genuineness of expression: the honor that he held high in his living.

Where Michael is precise and ordered—the quintessence of a manager—Nancy reflects her mother's artistic, imaginative spirit. A sensitive woman, Nancy feels things deeply and has an ability to empathize with a broad range of people. A short poem that she wrote not long ago captures something of the distinctive essences which she and Michael express, and how these essences blend together.

> In our pas de deux,
> broad-based and gentle,
> you step, toned and focused,
> I, free and fertile,
> in confluence of stride and turn,
> rippling through the days.

The two were married at Sunrise Ranch on May 3, 1967. Ross Marks was the best man and Diana de Winton the maid of honor. Martin escorted Nancy down the aisle; following a prayer by James Wellemeyer and a reading by Roger de Winton, he then proceeded to conduct the service.

Parents sometimes look to their offspring to accomplish what they themselves, perhaps, failed to accomplish. Martin reversed this approach, saying in effect, "I am responsible. It is up to me to do what is required, to set the example." Because of this, he opened the way for Michael and Nancy and many others to embody the same attitude—and find their own destiny and fulfillment in the process.

The couple flew from Denver to England the day after the wedding. During their seven-week honeymoon trip and speaking tour in Europe and the eastern United States and Canada they touched the hearts of many people, the first of them being a young Nigerian by the name of Odim Mkpa, who met them on their arrival in London and assisted them during their visit. Soon after arriving they held a meeting at their hotel attended by about twenty people; these all had some previous contact with the Emissary movement, but most had never met each other. Following Michael's talk Nancy sang some of her own folk songs, accompanying herself on the guitar.

The newlyweds held several other meetings in various parts of the country and visited members of Michael's family at Burghley House and elsewhere. While they were in London the Cecils invited Odim Mkpa and his girlfriend, Georgina Aspinal, for dinner—and it turned into an engagement party. Odim was in England taking a course in hotel management, while Georgina was a student nurse from Jamaica. On another occasion Michael and Nancy were invited to a popular London folk cellar where Nancy sang some more songs.

The twenty-year-old bride from 100 Mile House captured some of the essences of her visit to England through poetry:

London

Constant motion,
disorganized and amiable,
the tradition of black taxis
and brick rows,
swirls gradually into
the sensibility of friends—
laughing black eyes,
piercing green ones,
and blue with a twinkle.
Friends in London
one needs—
we had many.

Burghley

The 1500s,
Elegance and custom and etiquette.
But the ornate rooms were fresh,
looking onto the expanse of garden,
and were newly, artfully revived.
It was the Marquess and Marchioness—
they loved the place,
and welcomed us into it.
Shared conversation,
laughter,
a crackling fire—
and the Elizabethan Age became as
contemporary
as a snoozing dog, and a breezy day outside.

While a primary means for the expression of spirit has always been the spoken word, Martin was keenly aware of the value and importance of books. He had done his first writing in 1940. In the dark days of World War Two he wrote a series of stories, or lessons, for children, which were later published under the title *Child Light*. The stories centered on the theme of exploration, taking the child on a journey across the Desert of Patience to the Land of Heaven. His brother's three children, Davina, Gillian, and Angela, were staying with him at the time, having been sent out from England for safety. Martin used to gather them round on Sunday evenings along with Michael and read the latest episode he had written. Conrad O'Brien-ffrench illustrated the stories with drawings when they were published in booklet form in 1957.

Martin's first hardbound book, *Meditations on the Lord's*

Prayer, was published in 1967. In four extemporaneous addresses he brought a new perspective to the words that have been repeated by so many millions of sincere people over the centuries—with so little apparent result. Yet this beautiful prayer does find fulfillment, he affirmed, when human beings recognize the real responsibility it portrays. One aspect of this responsibility to which he alluded is the need to stop petitioning God to do something, and to recognize that God is waiting for human beings to do something. The necessary provision has already been made. Everything is present. All that is necessary is for human beings to recognize it and receive it.

"What a wonderful prayer this is. How complete and all-inclusive!" Martin declared. "The wonder and the beauty and the glory of that which is of God has been so obscured by the manipulations of the human mind, and yet it is so simple, so direct, straightforward, so perfect."

The following year, 1968, Martin published a second hardback book, *As of a Trumpet.* In succinct, universal language, he presented a poetic and comprehensive outline of truths that transcend any nationality and faith. The following is an extract:

Where are your primary values? Have your body, mind, and emotions convinced you that the values are properly in the world state that you now know? Let us consider.

Does the physical condition of your own body loom large to you? Do your material circumstances trouble you? Is it your experience that they determine your happiness or unhappiness? Perhaps changes in the members of your family, your friends, your business associates, are required to make you content? Is security found in money or position?

Are you easily hurt by what other people say or do? Are you quick to leap to your own defense? Do you feel that others should treat you differently or love you more? Does resentment, anger, or fear dominate your attitude and actions at times? Do you hate anyone?

Many more such questions could be asked. Be honest in your self-examination. If you say "Yes" to any of them, then to this extent your primary values are in the unreal state and you are identified with a false self.

If the nature of what you express in life is determined by the state of the world of your awareness, you cannot know your true Self. All you know is the experience of a puppet. Your only apparent hope for happiness and meaning depends upon your environment, most of which seems to be beyond your control. The people and circumstances out of the past and in the present encase you in a rigid personal bondage from which, try as you may, you cannot escape. What has come from the past makes the present seem hopeless and any bright expectations must be pinned to the future. All this is consequent upon the state of the unreal world, contained in the consciousness of the unreal self, but appearing to be an objective and very real environment.

It is futile for the unreal self to try to save itself from the devastating experience of its own self-created state. Its natural habitat is the unreal world and it can only exist therein. Each is dependent upon the other. Both need to pass away.

Following this passage, Martin proceeded to outline the beautiful qualities of the true Self which, when accepted into expression, bring forth newness and magic in living. The unreal world, he declared, must inevitably pass away, but this need not be ultimate disaster for humankind. If there are those on earth who have accepted and experienced the true

Self then a new world characterized by order and beauty will take form even as the old one disintegrates and falls away.

Pointing to the disintegration that was coming, Martin led a symposium at the University of British Columbia in February 1968, entitled "The Science of Survival." He used graphs to show how humanity was, as he put it, "slamming into a brick wall." First he dealt with the population explosion, showing how for centuries there was only a small increase in the numbers inhabiting the planet. By 1650 the population had doubled to an estimated five hundred million. From that point on the doubling occurred more and more quickly until by the year 2000 the population graph was going almost straight up, with an estimated population of seven billion, doubling again to fourteen billion by 2035. Martin pointed out that several other kinds of "explosions" were happening at the same time. Nuclear weaponry, for example, was becoming more and more sophisticated and powerful, well capable of turning the earth into a cinder. Explosions were also occurring in the fields of communications and information. To someone capable of standing back from the planet, it would appear that civilization itself was exploding.

The one thing that was not sharing in such an exponential increase, Martin suggested, was the moral and ethical state of human beings. He emphasized the need for exponential spiritual development that would match what was occurring materially.

Speaking at 100 Mile House shortly after the symposium, Martin pointed to the paradox of "The Hope of Hopelessness."

At our symposium in Vancouver, emphasis was placed upon the evident fact that the human race doesn't have much future. How little future there is on the basis of the present state has not yet really come home to the majority of people. We have some awareness of the subtlety of the conscious mind, the serpent as it has been called, and how it is able to rationalize and twist the facts, or such facts as are observed, to suit its own fancies. We could say that the conscious mind of man has certainly played the part of the deceiver. One of its principal deceptions is the idea that it can find a way of triumphing, in establishing its own idea of a world which possibly might be described as Utopian. Faced with the circumstances as they are in the world now, and as they may be projected into the future, there is still the idea that somehow the mind of man is going to be able to blunder through, should we say, and emerge successfully on the other side.

I suppose there is a word we might use to summarize the matter, and that is "hope." As long as there is hope in the usual sense, there is deception. Until the point is reached where in the awareness of the conscious mind the situation is absolutely hopeless, there is no true hope. Paradoxically, the moment this point is reached—and it is becoming more and more obvious—then immediately a basis for true hope appears. The answer may be experienced.

In July 1968 Lillian conducted the first Emissary music workshop at 100 Mile House; for many years she has been the guiding inspiration behind the musical development of the Emissary program. In April of the following year, another "first" occurred when nine Emissary centers across North America shared Martin's Easter morning service on Sunrise Ranch through a coast-to-coast telephone hookup. "The irresistible force of spirit working through many people over the years has brought this fulfillment about," he commented. "There is an awareness of the reality of what may be referred to as one body."

At a conference at 100 Mile House in November 1969 Martin set the stage for the decade to come. Many people, he said, had begun to recognize the awful situation facing humankind. What was essential now was to move away from the mere consciousness of hopelessness and acknowledge the beauty and perfection of the divine state, "because we are in it and exemplify it. The stress then begins to be placed more particularly upon the right state than upon the wrong state. This is fitting and right as we move into the 1970s."

Shortly before Christmas Martin traveled alone to England for a short visit with his family. With his brother and two sisters he attended his mother's ninetieth birthday celebration at Leyburn, Yorkshire. While he was in England he also participated in a meeting at the home of a young man named Rupert Maskell, who was providing leadership and agreement with him in that country.

The way in which Maskell connected with the Emissary program is typical of many who heard and acknowledged the spiritual invitation that Martin extended over the years.

Born in London in 1938, Rupert Maskell accompanied the rest of his family to South Africa in the summer of 1939 to visit his maternal grandmother, who had married into one of Natal's pioneer families. War broke out, and they all stayed on, settling in the Cape Province. After finishing his basic schooling in South Africa, Maskell returned to Europe in the mid-1950s. He apprenticed in the wine trade in France and Germany, then opted for the bright lights of London, where he joined a firm of auctioneers and estate agents. He was successful from a material standpoint, but couldn't settle for the kind of future that he saw laid out in front of him—a junior partner at thirty, senior partner at forty-five, retirement at sixty. Instead he emigrated in 1962 to New York where, he had heard, the "sky was the limit."

There, he was eventually hired by an international firm of

property brokers who were looking for someone to run their London office. After some training in real estate he returned to London with an excellent job—the envy of his friends. Troubling questions kept disturbing his peace of mind, however. Why was he doing what he was doing? What was the purpose? In 1967, in London, he met an old friend of his parents named Helen Turner. As he described his feelings of disenchantment with the world she told him of a place in British Columbia called 100 Mile House which she had visited recently. She had found a group of people living together in a remarkably harmonious and effective way, and knew that she would have to go back. Maskell listened to her story with interest, then forgot about it.

Six months later, while on business in New York, he met Turner again and asked her out for lunch. She brought two friends with her, Alan and Jean Hammond. Maskell liked them enormously, confiding at one point, "There's something about you—you've got something that I'd like to have too." The Hammonds, who were directing the New York Emissary center at the time, invited Maskell to attend a public meeting the following night, which he did. Finally, he was impelled to travel to 100 Mile House to meet Martin, and see for himself what was happening there. He was welcomed into the flow of the community, and whether helping to skin a moose or vacuum the floors, he felt at home immediately, knowing that it was a turning point in his life.

By the time Martin arrived in London at the close of 1969, Maskell had taken a course at Sunrise Ranch and was preparing to start a small communal group in London with the Mkpas (whom Martin had married at 100 Mile House in 1968).

Following his return to Canada from England Martin

commented: "I have traveled to England many times in the past, and while, over the years, there have been a few who have played their parts and kept the light burning there has never been an adequate pattern of agreement present in that far country; but this time there was. There is something strong and stable there, something to which I could go with confidence."

As the New Year, 1970, dawned, the same quality of stability and strength which Martin had found in England shone also in the eyes and the lives of many others who loved the spirit which he exemplified. There was a base from which the influence of that spirit might encircle the globe.

20.

LETTERS FOR LIVING

They wrote of their heartaches and fears, victories and joys. For more than thirty years Martin carried on a personal correspondence with people from all backgrounds and walks of life, assisting them in the processes of spiritual maturing, helping them to meet the challenges of life effectively. These were people who wondered why they were born into the earth, what the real nature of spiritual expression was, or how they could cope with problems and crises in their lives. Sometimes their letters glowed with honesty and openness. Sometimes the writer was confused or afraid, perhaps suspicious or even hostile. Martin received them all alike, in an attitude of thankfulness and love. That didn't mean he didn't call a spade a spade at times. But he never condemned anyone.

Nothing was more important to Martin than his mail. Regardless of other needs that might be present—even his own physical needs—the mail had to be handled, and handled properly. Those who looked to him for spiritual orientation did not compose some kind of homogeneous mush, in his eyes: through his letters, thoughtful and penetrative, he offered each one the opportunity of developing an intimate, individual connection with him.

During the 1970s Grace Van Duzen gathered excerpts from some of Martin's letters to be distributed along with the regular Emissary mailings under the title, *A Word of Grace*. Later, these excerpts were published in a series of seven booklets entitled *Letters for Living*. Covering many areas of living, the extracts were offered "in the conviction that these same problems and questionings articulate a cry in the hearts of multitudes on earth today."

Health was a subject of perennial concern. As with all the other areas that were brought to him, Martin encouraged people to look at the apparent problem, or lack, objectively —to identify with that which was able to fill the lack.

The potential of what is called cancer is present in everyone. The factors involved relate to life processes which may, for one reason or another, become unbalanced or out of rhythm. Then the so-called disease shows up.

Identified with the truth of your own being, the proper rhythms have opportunity to balance out in your physical body. Your concern, then, is simply to express your own true and beautiful self in the right handling of all the opportunities that present themselves to you. You are not, then, centered in a lack, nor do you give thanks for it. You are simply thankful for the abundant opportunities through which you may reveal the truth of yourself.

As all this is so for you, you are indeed useful in the true sense and may stick around for a while, as you put it. The fact of the matter is, in such a case, that regardless of what happens in the changing, external state, you are eternally on hand.

★ ★

As you indicate, the spiritual must be put first, which would then presumably put the physical in a secondary position. However, the fact of the matter is, when you do put spiritual things first the physical state of your own body is immediately seen as being of importance. It is in fact also spiritual but at a different level of vibratory manifestation.

It then becomes evident that from the spiritual standpoint you do need to take your physical behavior in hand, as this relates, amongst other things, to the health of your body. Clearly, such matters as proper diet, proper exercise, proper breathing, and proper rest, become of immediate concern. A consideration of such things on your part need not cause you to become a health nut, because this is only part of the total balanced spectrum of your living. If you neglect it, however, it would be impossible for you to experience true balance in living, which in turn would indicate a failure in spiritual expression.

I would encourage you, then, to apply whatever discipline is necessary to make possible the revelation of the beauty of your true spirit by reason of the proportion and grace of your physical form.

A note of skepticism sometimes permeated the letters which came to Martin's hand. But there is value, he suggested, in "honest skepticism."

In a recent letter to me you ask whether it is really necessary for a person to find another to lead and teach him. You then say, "Isn't that the purpose of the spirit?" In replying to this I would first ask another question. If human beings were capable of being led and taught directly by the spirit, why have they not let it happen? There have surely been many people in every generation who were just as sincere and earnest as we may have thought ourselves to be, and yet what an immense amount of confusion, contradiction, and conflict has nevertheless been engendered.

It is wise, as you imply, to approach the offer of spiritual leadership with caution. I have no quarrel with some honest skepticism on the part of anyone making an approach to this program. Those who swallow too quickly what is offered inevitably do so without any real understanding and they would probably turn in some other direction just as easily. As long as there are gullible people they will be taken advantage of by someone.

If you have been reading our literature and listening to tapes, you must recognize that you have never been asked to accept anything on blind faith. Certain fundamental principles are portrayed for you and you are invited to let them be proven out in your own living. If you do this you will know whether they are true or not. This is the only way by which anyone may come to know the truth. As this becomes your experience you find that you are moving with the spirit, but it took someone else who was moving with the spirit to direct your steps in the right way. Then you know it for yourself.

Money and sex, it is said, are the two areas which cause the most pain and confusion in human lives. Regarding money, Martin had this to say on two occasions:

One can easily pay too much attention to laying up treasures on earth, where moth and rust corrupt. There is a need to act sensibly with respect to any financial responsibilities one may have, but never let it dominate the scene. It is an absolute fact that the only true value and security rest in the treasures of heaven. The reality of these is revealed by one's own steadfast spiritual expression.

I had thought a little about gold and silver, whether in the form of bars or coins, but amongst other things the matter of storage poses a question. Thieves do break through and steal, and it all might prove to be ultimately worthless. You can't eat metal.

The point really is that money still has its uses, so let's use it.

★ ★

I would agree with this paragraph quoted from your letter: "My feeling at this time is to leave the money right where it is until there is sufficient insight as to how to handle this rightly."

There is a right use for money but a person must find sufficient spiritual maturity to know what that right use might be in any particular instance. The ideas of the human mind of an immature person will have very little bearing in the matter. The larger the sum of money involved, the greater the maturity required.

In order to mature, spiritual orientation must be consistently maintained over an adequate period of time. During the process, opportunity will be present to be a sensible and trustworthy steward of such money as may be available. If one proves faithfulness over small amounts, then presumably one may be properly trusted with large sums. But the point is to learn the right handling of finances when they don't amount to very much. Otherwise there will be failure and waste when larger sums appear.

It would seem that money is about the most prevalent lure to cause people to revert from spiritual orientation into material orientation. In such case there is no way by which money can be used rightly.

Maintain your spiritual orientation, have patience, and be willing to learn how the substance of the spirit, of whatever nature, may be used rightly.

Many who wrote to Martin over the years were having difficulty in the area of personal relationships. Nowhere was his accuracy and delicacy of aim more evident than when he approached this source of so much anguish in human lives. While he grew up with his own share of ignorance and inhibitions in this field, as his alignment with spirit clarified he began to discover the true beauty and purpose available to be known by men and women in their relatedness together, and so was able to assist others into that same discovery.

His letters on the subject carried sweetness and encouragement, allied with an emphasis on the need for balance and patience.

I received your two recent letters with thankfulness, rejoicing in your spirit of openness and trust.

The things of which you write loom large in your consciousness now but may be allowed to recede somewhat as time passes. By this I do not mean that what you now experience is not important or should somehow be suppressed, but as you continue to keep your integrity the immediate intensity may be allowed to balance itself out.

You mention balance, and the young man with whom you have been going has observed a loss of balance in girls who become "hooked" on sex. The hesitancy which you have both experienced in this connection is based in spiritual wisdom. There is a saying that fools rush in where angels fear to tread. All too many girls have spoiled something most precious and sacred in themselves by being in too much of a hurry.

The true relationship between a man and a woman only becomes possible when there is sufficient maturity on the part of each one. This maturity is an individual matter which can be prevented by too close an association with a member of the opposite sex too soon. The social pressure of one's peers to move too close too quickly is something which you no doubt feel most keenly, but the fact that you recognize that there is something not right in the popular way gives indication of your movement in the direction of greater maturity.

At your age there are actually many areas for the greater expression of yourself emerging to be rightly used. There is far more than this one aspect related to physical sex. Let your awareness broaden, so that your life energies are no longer channelled so exclusively into this one field. There are a multitude of other ways in which life reveals its real and balanced beauty.

★ ★

I understand what you are feeling. A sense of loneliness is quite natural in your circumstance at this present time. It may be somewhat uncomfortable but there is nothing wrong with it. In fact, it can and should be an essential part in the right unfoldment of your life.

By the same token, there is nothing wrong with the desire for close companionship in the external sense. However, this would not contribute to an anticipated fulfillment unless it is allowed to take form at the right time according to the true design. What you might envision in this regard is not necessarily what it should be.

As I mentioned before, the tuggings at your heart, as they occur in various ways, are valuable in bringing to pass the right outworking, provided that they are not allowed to control you. This can only be true to the extent that you maintain your centering in spirit. Fundamentally this means that you feel more deeply about being true to your highest vision than to your longing for external fulfillment. I am sure this is really your attitude. Don't deny your true heart's longing because the superficial one is apparently making such a clamor.

★ ★

I would assure you that what you are and have been experiencing can be of the greatest blessing. It may be uncomfortable for the moment but your manhood may be strengthened thereby.

Our circumstances are never entirely self-produced but the way we handle them and our attitudes toward them are always one's own responsibility. One may feel lonely, hurt, or whatever, but such feelings should never be claimed as though they were an indication of one's own true identity; they certainly are not.

Your manhood puts in an appearance to the extent that your behavior proclaims the strength of your own manhood rather than the weakness of your momentary feelings. No matter how fine a particular woman may be, never hand over the control of your life to her by reason of your feelings.

In another letter Martin spoke of the sacredness and joy that may rightly characterize physical intimacy.

In writing to me you opened a door through which a blessing may be released. This is so in any case but I am taking the opportunity to write this personal note to you.

It is true that sex should not be looked upon merely as a means of getting pleasure. However, the fact of pleasure should not be seen as though it were intrinsically wrong. It is in fact a proper and necessary part of the true purpose of sex.

I am sure you will understand that what I could say now in this letter could not, in any way, begin to cover all the creative ramifications of the true sex experience. All I really can say is that it should be approached in the recognition that here is the highest form of worship that a man and a woman can undertake in love and purity of heart. Through this union there may indeed be union with God by reason of which there is an intense generation of spiritual substance which ascends as incense and also may be permitted to permeate all levels of one's own consciousness and living. This substance is essential if one is to let one's light shine.

If this most intimate experience is kept sacred, then all departments of living will be included in that sacredness and the glowing radiance of the spirit will be apparent in all one's ways. In approaching the sex experience with your husband you are in fact offering yourself to spirit in an attitude of yielded love.

The "Myth of Progress" was the theme of this fiery epistle:

There have been sincere people in every generation, following out their particular spiritual insights for millennia, and yet the state of human nature on earth has remained virtually unchanged. Civilizations have come and gone, some of them beyond the range of present historical memory, some of them perhaps at least as magnificent as this one. When the civilization was on the rise it was invariably considered that progress in an evolutionary sense was being made. Some have thought of that evolution in material terms, others in spiritual, but then disintegration came and it was obvious that no progress at all had, in fact, been made. The whole spectrum of human consciousness, from the material extreme to the spiritual extreme, has been characterized by a state of self-deception. If this is to be transcended it can never be done by a rehash of the same old approaches that have been repeated down through the ages ad nauseam.

The transcendent experience must be unique in its nature.

Everyone grows old sooner or later, unlikely as the prospect seems when one is young. Martin devoted much time and care to encouraging and assisting those of advancing years.

It is not surprising that you feel a certain concern about growing older, as this has been the main preoccupation, consciously or unconsciously, of human beings ever since they lost consciousness of the truth. However, there is no necessity to be subject to such feelings, which merely derive from the unfortunate physical and mental state in which human beings have continued to flounder.

It would seem that I am somewhat older than you are, having advanced well into the senior citizen category. No one is exempt from what is likely to occur in the aging process, but you remain you regardless of that and may continue to reveal the true quality of yourself through all these circumstances. To the extent that this is your expression, that is what you will know and you need no longer be entangled emotionally in the state of your physical body. That is not you; it is merely your body.

As long as your spirit remains strong your living will continue to reflect it in ways that are fitting. Enjoy living now because you express to the highest and fullest the real quality of yourself into the circumstance of each succeeding moment. If you do this you will have neither time nor inclination to beg trouble from an imaginary future. You will love life too much and you will find yourself inseparably associated with it.

★ ★

The sense of depth which may be offered by a person of many earthly years is valuable as a balancing element for those who are younger, even as their vibrancy of life can be a blessing of uplift to the older ones. All are necessary in the collective form which increasingly makes possible the balanced expression of spirit on earth.

In this last short letter, Martin gave away the key to happiness:

It was good to hear from you.

The child is always intent upon doing what he enjoys. It is natural for a child to be self-centered in this way. The evidence of increasing maturity rests in the discovery that it is quite possible to enjoy whatever it is one does. The enjoyment comes because one pours oneself wholeheartedly into whatever it is that properly comes to hand for the doing.

The increased experience of life is thereby made possible, and life is an enjoyable experience—so there you have it!

Martin's letters, pithy and to the point, helped many hundreds of people to discover the happiness and fulfillment natural to the true experience of life.

21.

EAST MEETS WEST

Oh, East is East, and West is West,
    and never the twain shall meet,
Till Earth and Sky stand presently at God's
    great Judgment Seat;
But there is neither East nor West, border,
    nor breed, nor birth,
When two strong men stand face to face,
    though they come from the ends of the earth!

Those who know Kipling's famous words—"... and never the twain shall meet"—sometimes have not heard the lines that follow. But certainly East and West do meet, the distractions of nationality, breed, and birth do dissolve, when people come together on the basis of genuine strength of character—a shared love for the true quality of life. As the

1970s unfolded, it became evident that many all over the world longed for an experience of friendship and agreement that transcended racial and other barriers. Some of these men and women found, in Martin, and in those who were standing with him, the means to actualize that deep-felt longing. With integrity as its common meeting ground, the grouping of people for which Martin was responsible began to assume global proportions, reaching out with greater strength and effectiveness to Africa, and also establishing itself in Asian countries, in South America, and elsewhere. East does meet West—as there are those who heed the call of spirit.

In England sixty people gathered at a hotel in the heart of Westminster in July 1971 to share in the first Emissary symposium held in that country. "By the end of the afternoon even the stiffest upper lips had relaxed into warm smiles as people felt and saw the spirit of agreement at work," Rupert Maskell reported. The program included a tape of one of Martin's talks. Under Maskell's leadership regular meetings were also now being held in Italy, South Africa, Ghana, and Nigeria.

In the fall of 1972—through an unexpected turn of events—I became involved myself in the expansion which was occurring. I had been editing the 100 Mile House weekly newspaper for a number of years, and one day wrote a letter to other newspaper editors around the world, expressing some thoughts on the theme of spiritual values. To my great surprise the letter ended up being printed in over one hundred English-language newspapers in approximately thirty countries. It said, in part:

Since all attempts to arrive at material solutions have evidently failed, it is surely high time to consider placing true spiritual values, such as integrity and the sense of personal responsibility, at the very head of the list and letting the chips fall where they may. Such a course probably would not find favor with many of the established institutions, perhaps even including some which are religious in nature. But I am convinced it would provide the means for the emergence of a vitally-needed factor: a worldwide brotherhood of men and women united on the basis of a common dedication to the qualities of true manhood and womanhood, of their own inherent greatness. I would be very pleased to hear from any of your readers on this matter.

Having dispatched the letter to fellow editors in distant places I sat back and waited. The letter might not even be printed. Even if it was printed, would anyone be interested enough to reply? Three or four weeks went by, and I had almost given up on the experiment when a letter reached my mailbox from Hong Kong. A man had read the letter in the *South China Morning Post*. A few more days went by, until one memorable morning over fifty letters arrived from India and Sri Lanka. It was the beginning of a small avalanche. Hundreds of letters arrived from countries as far apart as India and Argentina, Australia and Japan. Because of the response my wife and I launched an international newsletter named *Integrity*. Following publication of a Spanish version of my initial letter in some newspapers in South America, we also started a Spanish edition—*Integridad*.

One of the main themes in the letters arriving at the little 100 Mile House post office from the far corners of the earth was the idea, "I thought I was all alone—that nobody else felt this way." In letter after letter, those who wrote expressed their delight to discover that this was not true—that, as one person put it, "I'm not crazy after all." An elderly gentleman in Osaka, Japan, stayed up until one in the morning to express his love and appreciation for life and for what he felt when he read the letter in his morning paper. A man in Calcutta said he was "struck dumb with wonder" when he read thoughts so close to his own. A rear admiral in Sri Lanka stated emphatically, "The desire and hope of men to be united is a posture of permanence." A secretary in Johannesburg said she had learned that every time she put integrity first, things always worked out for the best.

Such interest was encouraging because it showed that spirit was indeed on the move everywhere and that many people were awakening to its presence. Why else, as Martin remarked at the time, would anyone bother to answer such a letter? "The very fact that there has been this response is remarkable," he said in a talk. "First of all, it requires that somebody read the letter in the newspaper; then, having read it, that they write a letter. Now, what percentage of those who read the newspaper would read the letter? And what percentage of those who read newspapers would read that particular newspaper? And so it goes: a refining process, isn't it? Yet there are, relatively, so many who actually took pen in hand or placed their fingers on the typewriter to reply to the letter."

Martin had felt the need for a new assembly hall at Sunrise Ranch for some time. As the Emissary program continued to develop, more and more people were coming to the Ranch for courses and visits. He sensed that before long there would be many more.

During the spring and summer of 1973 he discussed various possibilities for a new facility with John Summerbell, who had built the Red Coach Inn at 100 Mile House. The latter had been a successful contractor in Toronto, but like Rupert Maskell and others, became dissatisfied with his success. While attending a Christian Fellowship meeting at his local YMCA, he met a real estate salesman who told him about a booklet he had read by a man named Lloyd Meeker. Within a year Summerbell was taking a class at Sunrise Ranch.

With his characteristic innocence and optimism, Summerbell expected that the new assembly hall would be orthodox in design and therefore relatively simple to build. Things turned out differently. One day, after a violent, 130 mph wind swept through Eden Valley, Martin invited him to his office and said he had reservations about erecting a building with orthodox vertical walls; such strong winds could be devastating, he feared.

"I think we should build a geodesic dome," Martin said.

Summerbell's jaw dropped. "A geo . . . what?" he asked, knowing nothing at all about this novel form of construction. However, he began to do some research into geodesic domes, while entertaining the secret hope, as he was to confide later, that the whole idea would just disappear.

During travels through Colorado, New Mexico, Arizona, and California, Summerbell learned that most of the domes in existence at that time were spherical, shaped like half a ball, and suffered from poor acoustics. The idea for such a structure began to fade. However, when he heard that Buckminster Fuller, the father of the geodesic dome, would attend a design conference in Aspen, Colorado, Summerbell decided to go so that he could see him and possibly discuss the acoustics problem. As it happened, one of the exhibitors had erected a small dome patterned after an ellipsoid rather than a ball. This dome had a much lower profile and its acoustics seemed to be fine.

Everything began to fall into place. Summerbell was able to catch a few moments with Fuller as he walked down a corridor. The famous inventor and philosopher assured him that a larger version of the ellipsoid dome which he had seen could be safely built as long as all cuts and angles were accurately done. It turned out that the people who had erected the ellipsoid at Aspen were able to provide a complete breakdown of the angles and lengths needed to implement Martin's design.

In August 1973 Martin led a group of one hundred and thirty men and women to a knoll at Sunrise Ranch which, it had been decided, would be the site for the new assembly hall. It was a dramatic moment. The annual Emissary conference was in progress, when people from many parts of the world come together at Sunrise to consider matters related to the Emissary program. As a fresh summer breeze caressed the hill Martin invited the assembly to hold hands and form a circle. Standing at the center of that circle, he then dedicated the site of the new building that would take form in the months and years to follow—its perimeter now outlined in living flesh.

The Dome, as it was called, embodied the love of many people. More than two hundred volunteers helped construct it, coming from many parts of North America and elsewhere. Buckminster Fuller's advice to be accurate was taken seriously; all the angles were cut within a 32nd of an inch. At one point during the spring of 1974 three crews worked round the clock for forty hours, endeavoring to pour the building's ferro-cement skin without a break. All the rest of the Ranch personnel were on hand to give encouragement, and Martin and Lillian helped serve tea to the crews.

In a geodesic dome, the structural supports comprise a network of wooden triangular struts that support and complement one other. If pressure comes on one area, every other strut in the roof goes to work to share the load. In a way, the Dome is held up by air—just as cellular interaction and compression holds up the branch of a tree and prevents it from breaking in a storm. Moreover, each triangle is individual—they are not repetitions of one another. Adding to the flexibility of the structure, the various struts are joined not by nails or bolts but by metal straps that allow for movement. Someone discovered, after the building was finished, that its vertical axis correlates with the so-called "golden cut," or "golden section," which since the time of ancient Greece has been a guideline for artists and architects in formulating aesthetically pleasing proportions.

With seating for 464 arranged in a slanting theater format, and unique acoustics, the Dome is highly functional. But it is more than that. It has a noble, spacious atmosphere—"a natural dignity and order," as Martin once described it. Large windows provide a close sense of connection with the rich pastoral setting outside.

When Martin gave his first address in the Dome (in May 1976) two birds gained access to the not-quite-completed building and, from their lofty perches, added a sweet accompaniment. The presentation was informal, in keeping with the informal, unfinished condition of the interior, but for those who were present it was an awe-inspiring event that no amount of conventional pomp and circumstance could equal. One sensed that it was more than the useful inception of a building: it was a step in the fulfillment of a design which had

Lloyd Arthur Meeker (drawing by Conrad O'Brien-ffrench). Right, Martin and Lillian soon after their marriage in September 1954. Below, with Lillian and Grace Van Duzen, New Year's, 1960.

Preparing for take-off. Ross Marks in the pilot's seat. Right, at the controls of a "Cat" – a favorite activity.

A light-hearted moment at Martin and Lillian's twenty-fifth wedding anniversary.

Above, Martin's immediate family. From left, Helen and Luke Vorstermans with Benjamin (on lap) and Rebecca, standing at front; Michael and Nancy Exeter with Angela (standing behind Rebecca) and Anthony, seated; Lloyd and Paula Meeker, at back, with Lloyd James; and Peter and Marina Castonguay, with Majessa (on Anthony's lap) and Dylan.

Above, from left, Michael, Nancy, Marina, Helen, and Lloyd. Left, Martin with cousin George Cecil at Biltmore House, North Carolina. The mansion was built around the turn of the century by George Vanderbilt.

Father and son.

Father and daughter.
Left, with son-in-law Peter
Castonguay, manager of Bridge
Creek Ranch.

Nancy and Michael with friends during visit to Tokyo, Japan, November 1987. Below, at home with Angela and Anthony.

The Lodge, built by Martin in the early 1930s, is still a symbol of welcome to all who visit.

Interior of the Dome, Sunrise. Officiating at tree-planting ceremony.

"His heart never ruled, neither did his mind. I always found his beautiful spirit ruled the situation . . . " —Lillian, speaking of the man she married.

been established long ago, and of which the Dome was merely a visible symbol.

Martin always equated spiritual expression with common sense. I doubt that the Lodge community, for example, would have lasted as long as it has if it was not based in something sensible and practical. This practicality springs forth from people, from the consistent expression of such qualities as integrity, trustworthiness, and love. What is true of the Unit at 100 Mile House is true of other groupings and individuals within the Emissary program. To the degree that those concerned have set aside traditional human habits such as accusation, blame, and criticism, accepting responsibility for their own behavior, they provide space in which the living essences of spirit can take form. Those essences, which include the beauty of design and control, have always been present, but they have no meaning until they are expressed.

As the 1970s unfolded, this process whereby the invisible design of spirit becomes visible took an interesting turn. It became evident that an inner core of agreement had formed around Martin—men and women who shared his own unequivocal passion for the clear expression of spirit. This group—now known as "Executive Council"—met for the first time in March 1974 at Theodore, Alabama. Martin, Lillian, Grace, Michael and Nancy Cecil, and Bill and Donna Bahan, shared in the sessions.

He had looked to this event for a long time. It was not something that he could force, however, or even hasten. It depended upon the growth and maturing of the whole Emissary body, for the Executive Council is only a part of that formation, just as the head is only part of an individual physical body. Speaking at Sunrise Ranch immediately after the gathering at Theodore, Martin used the analogy of a trumpet to portray what was happening. For a trumpet to produce sound, he pointed out, there must obviously be someone present to play the instrument, but there must also be a trumpet. Essentially the trumpet is composed of two parts: the mouthpiece and the body of the trumpet itself. The gathering at Theodore provided the "mouthpiece," which when attached to the body of the trumpet makes it possible for the trumpeter to produce sound. The trumpeter, he suggested, is God—the power of spirit. But while it might be said that there is music in the soul of the trumpeter, that music remains unknown until the "trumpet" is available.

Nancy Exeter (formerly Cecil) speaks of that first gathering at Theodore and of her own experience of the Executive Council:

We were stiff that first time and on tenterhooks about what was expected of us. I remember Martin saying something about how the conversation could flow naturally. While those first steps had to be taken, the difference now is very striking. There is indeed something flowing—a broader base in our togetherness, in which what needs to unfold can do so more simply and easily. I suppose it could seem from the outside as if we gather and make all kinds of arbitrary decisions, but it isn't that at all. The Executive Council doesn't make decisions. We simply accommodate, as well as we may, what should come through us. Others have to be able to accommodate what should come through them. In that sense it's a completely democratic process. There's an impulse of spirit coming through and we all have to be able to give voice to it.

I remember one gathering where we joked a number of times about how there were a lot of balls up in the air and we just had to be content to leave them there and let something keep working. You can't just say, "Well this looks like a good idea," because if you do that and push it through it may turn out later that it wasn't a good idea at all. So we will discuss something for a while and then maybe a day or two later we'll come back to that topic and see if any of the balls have come down! It's not a matter of everyone contributing their ideas and then seeing how the ideas fit together. It's a matter of contributing what is in one's heart and mind to contribute and then letting it grow so that the thing can develop of its own accord; it isn't being pushed around by anyone.

As life has moved on, there have been changes to the Executive Council. Its number, for example, has grown from the original seven to twelve. The purpose of the grouping remains the same, however—to provide a "mouthpiece" so that the music in the soul of the trumpeter may continue to sound and fill the earth.

A Methodist minister named George Emery, with his wife Joelle, began to play an active role in the development of the Emissary movement in the 1970s, both in North America and further afield. The Emerys first met Martin during a visit to Sunrise Ranch one Easter. Emery likes to tell the story of how after being "gloriously turned on" by Martin's words he visited some people at the Ranch and was disconcerted at seeing pictures of Martin in some of the homes. He wondered if it was a hero cult, though he did not feel this coming from Martin himself.

In his talk that same evening—as if answering Emery's concern—Martin commented, about halfway through, "You know, there are people who still try to put me on a pedestal, and I wish they wouldn't." At the end of the address Richard Cable, a pioneer resident of the Ranch and husband of Kathy Meeker's sister, Billie, spoke up and said, "Martin, you are on a pedestal. You just want us to come up and join you there."

"I've told that story many times," Emery recalls, "because I think that was the beautiful thing about Martin. Many people in the Christian Church go to an extreme of one sort or another. Either they put Jesus on a pedestal and leave him there so that there is a separation, or they bring him down to a humanistic level. What I loved about Martin was that he said, 'Come and stand with me and accept your true identity, revealing the spirit on earth, because if I can do it you can do it, and what is true of me is true of you too.' He never said, 'Follow me from a distance,' but rather, 'Come and stand with me in friendship so that together we can get the job done.'"

With the help of imaginative fund-raising events held by different people in the Emissary family—and a $1500 donation from Martin—I made my first visit to *Integrity* readers in India and Sri Lanka in 1975, accompanied by Achal Bedi. Originally from Jamshedpur, in northern India, Bedi came to the United States after graduating from university and found work as an electronics engineer, settling in California.

It was exciting to finally meet those with whom I had been corresponding for two or three years. The sense of friendship and agreement was immediate. Despite the external poverty, the people impressed me with their natural warmth and sweetness.

Martin kept a close eye on our progress. One morning, as we ate breakfast on the verandah of the Galle Face Hotel in Colombo, Sri Lanka, an envelope arrived with a copy of his latest talk, in which he had zeroed in upon the plight of countries such as India. "I wonder how anyone, looking at the state of mankind, could not be moved with compassion concerning those millions of human beings on earth who are completely without hope," Martin began.

I think this rather emphasizes a sense of revulsion with respect to those other human beings, ourselves perhaps included, who have had the opportunity of awakening to true purpose but have rejected it. It is those people living in the more affluent parts of the world, heretofore, who have had the greatest opportunity. All that has happened, so often, is that a greater attitude of greed has been engendered. How vital it is that awakening to the real need and being moved with compassion, we set aside the usual personal motivations which have become quite acceptable in this part of the world. What petty things have been raised to the position of false gods in our own day-by-day living. What we might perhaps think we want, and what we might perhaps think we don't want, have almost invariably dominated the consciousness of people and have lingered on in our own experience. What difference does it make whether, as individuals, we imagine we are satisfied or dissatisfied, comfortable or uncomfortable? What difference does it make? Have things come to a sufficient pass to convince us that all such pettiness is utterly meaningless? It doesn't warrant a thought.

The words paint Martin's character so vividly. He was totally unconcerned with personal wants. As it happened, he did not have opportunity to travel to India or other developing nations; but I know that because of his breadth of love and compassion, and because in his own personal life he scorned waste or excess, in his heart he was able to accommodate and welcome people of all races, classes, and cultures.

22.

THE TRUE STEWARD

In his book *Archy and Mehitabel,* written in 1927, American poet and playwright Don Marquis wrote: "it wont be long now it wont be long till earth is barren as the moon and sapless as a mumbled bone" (Archy, being a cockroach, was unable to punch in capitals or punctuation on the typewriter). What would Marquis say now, after another sixty years or more of human greed, depredation, and meddling?

An attitude of love and respect for the earth was a central element of Martin's approach toward life. He too rued the awful destruction wrought upon the earth, but took the pragmatic view that having been raped, it cannot be "unraped." What a person can do, he declared, and must do, is to relinquish the terrible fallacy that the earth belongs to humankind, to do with as they will. Because—unthinkingly, for the most part—human beings everywhere accept this fallacy as being true, things have reached their present pass, with the planet indeed becoming "sapless as a mumbled bone."

"The earth is the Lord's, and the fulness thereof; the world, and they that dwell therein." Martin loved these words of the psalmist, pointing to a larger design and purpose within which the planet and all creation are rightly contained. Such a notion does not sit well with the human ego, of course, which besides thinking of itself as the owner, also imagines that it is quite capable of managing planet earth effectively, thank you. Does it look like it? Martin would ask wryly. And if it doesn't look like it, does not common sense demand a change of approach?

Martin emphasized in many different ways that the only valid starting point for such a change is the individual. From this standpoint the fate of the world depends upon individuals—those who are honest and humble enough to admit that in trying to run this once proud and beautiful estate according to their own ideas and determinations, human beings have brought the earth to ruin. Such an acknowledgment opens the way for re-alignment with the larger operational whole—without which anything that anyone may do inevitably courts disaster. There is a vital task for men and women to play in caring for the planet. But separate from spirit, Martin insisted, dependent upon their own limited vision, driven by the fear and greed inherent in their isolated state, human beings lack perspective. They are unable to see the ramifications of what they do. The more admirable their goals the greater, as a rule, the ill effects that sooner or later appear. Once he quoted an investigation undertaken in the United States aimed at solving the problems of a large city.

A variety of data was fed into a computer, but the answer it came up with was that *anything* that was done would only make things worse.

This emphasis upon the need to respect the larger whole, so that true understanding and care of the earth is possible, is also a central theme of Martin's son, Michael. "Because we seem to have lost the ability to discern what fits in the overall design, all our good ideas turn out badly," says Michael in his book, *My World, My Responsibility*. He tells of a Frenchman who a century ago brought a small creature called the gypsy moth to the United States. This man wanted to cross the gypsy moth with the silkworm moth to produce a more hardy, disease-resistant strain of silkworm for his native France. Wrote Michael: "Unfortunately, while he was working on this some of the caterpillars got away, and the descendants of these creatures now pillage the northeastern United States and regularly demolish the leaves of some five million acres of trees. And this was such a *good* idea."

Martin first began to learn about stewardship of the earth in 1930 when he arrived at 100 Mile House. In that situation he was keenly aware that the Bridge Creek Ranch did not belong to him. He was managing it on behalf of his father. He had to respect his father's wishes, his father's vision for the property. As the years went by, Martin realized that the same principle was true on the larger scale: yes, he and other human beings were responsible for the care of the earth—but as stewards of the "grand design," not as isolated individuals each promoting their own views and opinions.

The Emissary communities which developed at 100 Mile House, Sunrise Ranch, and Green Pastures, in New England, all provided opportunity to learn true stewardship. In

1977 a further opportunity arose when Emissaries purchased a property named Glen Ivy.

Two hundred years ago the Temescal valley of California was inhabited only by the Luiseno Indians, a peace-loving people who appreciated the valley's rich mineral resources and the hot and cold springs that provided such a rich and constant source of water. The name *Temescal* was an Aztec word meaning "sweat-house," and referred to the mud saunas which these Native Americans built around the hot springs.

In the late 1800s a Captain Wheaton Sayward decided the hot springs of the Temescal valley would benefit his wife's circulatory ailments, so he brought her there, along with his seven children, and built a sturdy adobe house. It was the beginning of what would later become known as Glen Ivy Hot Springs, a country inn with restaurant and bathhouse that did a good business through the Prohibition and Depression years and flourished even more after World War Two. The popularity of such resorts waned, however, during the inflationary 1960s, and rising labor costs made maintenance impractical. Vandalism and a flood took their toll. By 1977 the property was up for sale at about half its 1964 selling price.

John and Pam Gray, Emissary coordinators in the southwest, had recognized for some time the need for a larger headquarters for the region. Late in 1976 two friends of the Grays visited Glen Ivy to look at a used car which had been advertised. The car was not suitable—but Glen Ivy, it soon became apparent, was. Despite its rundown state it offered the necessary facilities and was strategically located near the town of Corona, about an hour's drive southeast from Los Angeles. The Emissaries began negotiating purchase of the existing lease option. Martin visited Glen Ivy in January 1977, dedicating the new property in an address later entitled, "The Oneness of Heaven and Earth."

Here we are on this particular plot of ground, scarcely discernible on a map of California, let alone on a map of the world. And yet, it is a part of the earth. And to us, we begin to see it as being sacred ground, a patch of the earth, so to speak, which is being deliberately returned to the control and use of spirit. Exactly what that means we need to see increasingly clearly. But for the moment, I wish to emphasize the point that when you drove in here, you were not particularly aware of the moment when you crossed the line which delineates this property. It is an invisible line, arbitrarily established by human beings who think to divide the earth for their own purposes. Insofar as the actual fact of the matter is concerned, there are no such delineating lines on the surface of the earth. There may be distinctions between one part of the earth and another—the shoreline, for instance, with water on one side and earth on the other; or there may be rivers and mountains, forests and plains—but we are aware of the fact that the earth, and everything in it, belongs to spirit. This is the fact of the matter, whatever human beings may think about it and however they may attempt to divide the land for gain. There are various geographical positionings on the surface of the earth which once more are in the process of being restored to the purposes of the whole. We see this particular property in that light.

Many miles away from Glen Ivy across the Pacific, Lawrence Sullivan, a Melbourne carpenter, had maintained a close agreement with Lloyd Meeker and Martin ever since 1937, when someone gave him a copy of Meeker's booklet, "Lighting the Way in You." Sullivan visited 100 Mile House in later years, spending several months at the Lodge community. He became a good friend of Martin, who fondly called him "Lawrence of Australia," and also met Meeker. In March 1977, after being the only Emissary in Australia for forty years, Sullivan welcomed Dr. Paul Blythe, who emigrated from Canada with his family to take up a position as a lecturer at an institute in Adelaide.

Canada, that unobtrusive and generally well-behaved nation, nearly split apart at the beginning of the 1970s when the move for Quebec independence escalated into violence. Strong action by Prime Minister Trudeau in mobilizing troops and imposing the War Measures Act contained a crisis that many feared would erupt into revolution or anarchy. However, when the people of Quebec elected a separatist government in 1976, it prompted the federal government to launch a "Task Force on Canadian Unity," with eight commissioners from across the country. Trudeau invited Mayor Ross Marks of 100 Mile House to represent British Columbia and the Yukon Territory.

The work of the Task Force lasted two years and included a series of public hearings in fifteen cities across Canada, in addition to extensive travel within the commissioners' own areas. The commissioners submitted a lengthy report to Ottawa with their findings and recommendations. While it is not easy to measure the success of the Task Force, the chances are that it played a useful and significant role in improving relations between French- and English-speaking Canada. Certainly it galvanized public concern and interest around the theme of unity, providing a forum where people could air their perceptions and ideas. The Task Force encouraged an atmosphere of tolerance and understanding, which was no doubt helpful; and several of its recommendations have been implemented over the years in one way or another—although not always with credit to the source. "I think it's fair to say that we started the movement toward greater understanding," comments Marks. "Sometimes, when a new measure was introduced later, I would chuckle to myself and say, 'Yes, we said that in '78 or '79.'"

"The implications of what the Task Force is undertaking reach further than the borders of Canada," reported *Newslight*. "Because of Canada's role as a mediator in world affairs, people are beginning to see what can be exemplified for the world by the wise handling of her internal problems—if Canada can't get her own act together, then what advice can she give to anybody else? In the view of those on the Task Force what is needed on the part of people is a generosity of spirit and an openness of heart so that there is a willingness to move away from parochial, self-centered attitudes."

From Martin's standpoint, participation on the Task Force gave Marks an opportunity to offer a stable, balancing influence into a difficult and divisive debate. As Martin pointed out, there was no answer to the question of unity so long as people insisted upon trying to work things out at the level of their differences. Only when there was a willingness to transcend those differences and let something happen at a higher level—that of spiritual expression, where oneness is already a fact—could any resolution occur.

It takes courage to do this: to set grievances and fears aside and rise up to higher ground. In November 1977 the world witnessed a remarkable example of such courage when Egyptian President Anwar Sadat traveled to Jerusalem to meet with Prime Minister Begin. It was the first time since the creation of Israel in 1948 that an Egyptian leader had met with an Israeli leader on Israeli soil. It may have been the first meeting of two such heads of state on a peaceful basis in Jerusalem since the Queen of Sheba visited there from Egypt three thousand years ago.

According to his reminiscences, Sadat awakened to his spiritual nature while he was imprisoned in a British jail. In his book, *In Search of Identity,* the Egyptian leader stated, "In the complete solitude of Cell 54, the only way in which I could break my loneliness was, paradoxically, to seek the companionship of that inner entity I call 'self.'" Another paradox: far from being embittered by his imprisonment Sadat was ennobled. As he commented, "I was able to transcend the confines of time and place. Spatially, I did not live in a four-walled cell but in the entire universe. Time ceased to exist once my heart was taken over by the love of the Lord of all Creation." Pondering these words reminds me of an analogy that Martin sometimes used. When you squeeze a bar of soap in your hand, the soap will go up or down, even though the pressure is the same in both instances. In life, when a squeeze comes, the same principle applies: one can allow the pressure to push one up just as easily as down.

Martin saw Sadat's visit to Jerusalem as a vivid example of what can happen when there is a willingness to rise above the divisive human nature state.

There was a willingness to acknowledge a common spirit, and on the basis of this common spirit the two men, Sadat and Begin, found a sense of friendship, shall we call it, a touching of a condition of oneness. Because of this most people, of goodwill at least, who observed the situation felt a sense of uplift, the opening of a door. Here was something beginning to move. That particular situation in the Middle East has been locked into a structured condition for a long time, but now something is beginning to move because of the fact that the irreconcilable state was transcended momentarily. Human beings have very little experience of what can happen when it is transcended, so the usual habit is simply to revert back down again and try to reconcile the irreconcilable, which is a futile endeavor. It is useless. No matter how many conferences may be held and who attends the conferences and what they talk about and what they claim to agree on, it has not changed the state of irreconcilability at all, because it is inherent in human nature itself; so it is still there, nothing has happened. The only thing that brings about a change is when the state is transcended at the level of spirit. And if there is the experience of this transcendent unity and that is held, something will happen with respect to those factors which previously were held in an irreconcilable condition. The transcendent spirit is the only basis for unification of that particular pattern in the Middle East, which incidentally is not isolated; it is something that is prevalent throughout the whole body of humanity but it comes to focus there. We had better not look at that situation as though it was isolated from us. It is very close to us, as close as our own behavior.

Martin had kept a close eye for many years upon South Africa, not condoning apartheid, yet emphasizing the destructive effects of attitudes of blame and condemnation, no matter how justified they may seem to be. The human tendency, he would point out, is to look for a scapegoat. But what is really required is to face the discrepancies or misalignment present in one's own living of life.

The Emissary movement had been active in southern Africa for a number of years. In March 1978, seven years after Rupert Maskell first began holding meetings in Cape Town, Martin decided the time was ripe to visit the area. In talks in Cape Town, Kloof, Johannesburg, and Salisbury (now Harare), he suggested that heaven is known not because of a mental idea or concept but because of an actual flow of spirit in living. When that is so, a person is to that extent in heaven —in spirit—and the realm of form, while still present, no longer carries so much weight. Martin correlated the invisible heaven with the realm of cause: as there are those who learn to abide consistently in the place of true cause, true effects inevitably appear. His words touched the hearts of many, bringing assurance and encouragement and a new sense of trust in life.

The itinerary included an outing in a tuna-fishing boat off the spectacular Cape peninsula and an overnight stay at Hluhluwe, the country's oldest game reserve, home of the white rhino, buffalo, wildebeest, and other animals. Martin and Lillian and those who were traveling with them (including Bill and Donna Bahan) woke at dawn to see a herd of giraffe lumbering along the distant horizon in silhouette. Another highlight was a visit to the Zulu village of Empangeni.

After a stop at Johannesburg Martin continued on to Bulawayo and Salisbury, where he gave his first Easter address off the North American continent. Emphasizing the tremendous need for new vision and understanding, he remarked that Christianity has been a useful reminder of the truth down through the years, but has now become so encrusted with human ideas and opinions that it is seen as being the truth itself, which it is not. It is a reminder. To know the truth one has to express the truth.

Ockie Oosthuizen, of what is now Harare, Zimbabwe, was one of those who met Martin for the first time on his tour. "Your coming to Rhodesia has left me with such a deep sense of gratitude," he wrote in a letter to Martin. "Although we have only just come into the 'fold' we know that you have provided the point of focus and the steadiness over the years and I believe it to be important that we realize this and offer our thanks to you and those around you."

Oosthuizen and the others with him were indeed "new." Only nine months had elapsed since the Emerys spent an action-filled three weeks in Rhodesia, providing the first Emissary representation there. Amongst other things, the Emerys met with then Prime Minister Ian Smith and Bishop Muzorewa, visited Victoria Falls with an army escort, and gave a weekend seminar before a racially mixed group with African music as entertainment.

Arriving back at London's Heathrow airport, Martin and Lillian traveled to Burghley House to spend some time with Martin's brother David and David's wife, Diana. While at Burghley Martin also renewed his friendship with some longtime members of the staff, such as Hannah Atkinson, the Burghley housekeeper for many years. Martin had kept a close connection with his family ever since leaving England, writing regularly and visiting whenever he had the opportunity. On one visit, Lillian recalls, she ate chestnut pudding and jugged hare for the first time in her life, prepared by Martin's older sister, Winifred. "This was shortly after Martin and I were married," said Lillian. "Winifred is a wonderful cook and she prepared the meal herself to celebrate our visit. Martin always felt a special love for his sisters, and corresponded with them right up until his passing."

Returning to London from Burghley, Martin spoke before a large audience from many points on the continent as well as Britain. Bill Bahan had prepared the way for the event with a public talk the previous evening. Noting that many people were deeply worried about the future of the planet, Martin declared, "It is all going to come out exactly the way it should, if we play our part correctly. This is our responsibility, for we are all part of one thing, one spirit."

Not long after the completion of the tour, Emissaries in South Africa leased the Hohenort Hotel, a one-time family mansion in the heart of the Cape Town suburb of Constantia, as a business venture and headquarters. The word Hohenort is German for "high place."

Returning to 100 Mile House from Sunrise Ranch in November 1978, Martin received word of the passing of his longtime friend, George Shears. "This event was not unanticipated," he commented later. "George himself has had it in mind for a while. In fact he indicated that if anyone should find his body he should be the first to be informed! Here indeed was a prince of the kingdom, a man of stature. I am not merely referring to his considerable height in the physical sense but to a spiritual stature. Over many years a consistent steadfastness, a consistently right and beautiful attitude, has characterized this man." Such words sum up what is required of anyone who would be a steward of love and light for our planet.

23.

THE LIVING PROOF

It was not so much Martin's words that drew people from the four corners of the earth to share the action of spirit with him—necessary though the words were. It was what lay behind the words; his living. He took the everyday little circumstances that most people pass by as mundane, a waste of time, or a drudge, and filled them with the golden substance of love. In a way he was an alchemist, turning all he touched into gold. When I think about his life, it condenses in my mind's eye to the simplest of scenes: I see him sitting in his chair doing a crossword puzzle; picking up a piece of scrap paper from the road; or, to Lillian's consternation, staying up until 1:30 in the morning to fix a broken lamp. Paradoxically, what he happened to be doing in the moment, while it was always important to him, was never the primary thing. Whether he was tying his shoelace or addressing hundreds of people, his primary interest and concern was to use the moment at hand to express a clear spirit. Because of this, his living all came from the same place—a place of quietness and love; of respect for life's natural order and rhythm; and of humor.

Once a young Irishwoman visiting 100 Mile House noticed that a crab apple tree in Martin's front yard needed pruning. Being a trained botanist, she offered to take off the sucker growth and give the tree some shape. Martin accepted the offer and watched with interest from his living room window as, saw in hand, she climbed the tree and went to work. Very shortly, despite all her years of experience, the young colleen started sawing a branch in the wrong direction and cut her finger. It was a deep cut, and although she tried to apply pressure and lick the blood away as surreptitiously as possible, Martin saw what had happened and sped to the front door.

"Come here, young lady, and let's take a look at that," he commanded. Gently but firmly he applied first aid, commenting on the depth of the wound.

"I was going in the wrong direction," she admitted sheepishly. In a few moments the finger was bandaged and she was able to finish the job. A day or two later she went to see Martin to say good-bye.

"Ah, sure enough," he said, his eyes alight with fun. "Did you leave me a tree out there?"

"And where do you think I'd be putting it?" she replied, matching his Irish lilt.

"Ah, you have given it a haircut. But it will grow back," he said—and sent her off to Ireland with a bandage and a smile.

Martin could not abide waste. I came close to committing an awful blunder when in the course of picking up rubbish, I nearly hauled away an old sprinkler which he had left lying near his dustbin. Never mind that it looked like a museum piece and was held together chiefly by baling wire. It was still functional. It was *not,* he informed me clearly, garbage. This antipathy to waste characterized all of his living, including his use of words. He cared for words in the same way he cared for everything else, using them sparingly and accurately, and they carried weight because of it. He was also sparing in the way he distributed praise or compliments. He did extend a compliment once in a while. But basically he took the view that it should be natural for human beings to express their finest qualities—so why praise them for it?

For many years, Martin operated on the "firing line," giving public talks and seminars and conducting art of living courses. The nature and pace of his activities changed somewhat as he grew older, though many of the basic ingredients remained, such as his mail and telephone calls from far and wide. As the spiritual focus for all that was unfolding in the Emissary program, he necessarily kept an eye on a wide range of happenings. This included attending various meetings, adding his voice when needed. When it looked as if the Emissary board at 100 Mile House was in danger of burying itself under too many committees, he cleared his throat and made a pithy comment about unnecessary bureaucracy and red tape. What is required, he said, is for one person to take responsibility and do what needs to be done. Once, at a gathering in the eastern United States, someone complained that a certain group of Emissaries was being held back because their coordinator was "sitting on them." Martin defused what could have been a heavy and fruitless discussion. "When they're ready they'll hatch!" he remarked with a broad grin.

Martin was not easily distracted once he had embarked upon some task or other, which could be awkward, of course, if someone wanted to consult with him. One day he was cleaning his fish pond when his son Michael came along to see him. Naturally, Michael found himself joining in the work at hand while waiting for a chance to discuss what was on his mind; unfortunately, however, he slipped on the edge of the pond and fell in. As Michael recounted later, "I never did get to talk to him, having trudged home dripping. But that was characteristic of him —if there is something to be done, well, you do it and you don't think about how this is going to affect you, whether it's going to be convenient or becoming to your 'special and unique character.' You just do it and trust life."

Drafting and architecture were keen interests, and people often consulted Martin when a new building was being considered. The Dome, at Sunrise, which he designed and saw through to completion, was his largest project, but he provided drawings for many other buildings from meeting halls to dining rooms to private homes. Gardening was another favorite activity; the shrubbery, lawns, and flower beds at his home at 100 Mile House give rich evidence of his loving care over many years. "I remember the time when he went out and did a little forestry beyond our bushes there, amongst the aspen," Lillian recalls. "He carefully thinned them out so that each tree had room to grow naturally and easily. It's just beautiful there now, where he did some cutting away of what wasn't needed."

Exercise became a customary part of his later life. While he was a working rancher, there was no need to bother—it came with the job. As his life became more sedentary, he began to include an hour or so of jogging or walking. Once, when he was meeting with a small group, rain prevented any outdoor exercise, so he and the others ran up and down the stairs of the house. Following construction of a tennis court at both 100 Mile House and Sunrise, he often enjoyed a good game of doubles, entering into the fray with his usual wholehearted enthusiasm. His satisfaction, when he made a good shot, was evident. His anguish, when he made a bad one, was sometimes even more evident, as he raised an exaggerated, tortured groan to the heavens. But tennis was not an end in itself, much as he enjoyed it. He got out on the court because it was good exercise and, like cross-country skiing, helped his reflexes and balance.

I never saw Martin disturbed. I saw him act decisively and quickly when needed; but his inner calm remained unshaken —"not a leaf, as it were, astir upon the tree," to quote the North American Indian, Ohiyesa, speaking of the first American. Michael Cecil and I were driving back to 100 Mile House from another town when we heard, over the car radio, the news of President Kennedy's assassination. The event was causing shock waves throughout much of the world. But when we entered Martin's living room upon our return we found him sitting quietly in his chair, a half-completed crossword puzzle on his lap; his calm, reassuring smile carried as much impact as the shooting. It was obvious that whatever terrible things might be happening on earth, at the level of spirit all was very well indeed.

This internal stillness and trust revealed itself in Martin's approach to travel. Sometimes the factors related to a proposed trip refused to fall into place until the eleventh hour. If that was what was necessary to accommodate the movement of spirit, it was fine by him. His advice on this matter: "I'd get pressured a bit sometimes by various ones to decide, but you can't decide. It's not a matter of deciding; it's a matter of becoming aware of what the direction is. And if you're not aware of the direction, well don't go off in *all* directions, or don't arbitrarily make a direction. Let it reveal itself. And if it so happens that it doesn't reveal itself until about an hour before you need to leave, depending on the situation of course, that's the way it is. I don't think it's frustrating unless you make it so."

During Martin's earlier years in the Cariboo, young Ian Morrison was helping brand cattle at the Bridge Creek Ranch corrals one day when he cut the back of his hand with a knife. The cut went deep into the vein, posing a critical challenge in remote country far from doctors or hospitals. Alex Morrison, the ranch foreman, brought his then thirteen-year-old son back home to the 105. By the time they arrived, the boy was white with loss of blood. Neither Morrison nor his wife could stop the bleeding, and instinctively Harriet Morrison phoned Martin. "He always knew what to do and you knew you could depend upon him," she recalls. "He hurried from 100 Mile with a first aid kit, put a tourniquet on, and stopped the bleeding."

Another time Mrs. Morrison was working at a check-out counter at the 100 Mile Food Centre when Martin came in and bought some cigarettes. She was puzzled, because she knew that neither he nor Lillian smoked; later she learned he had sent the cigarettes to a ranchhand who had been locked in jail for a misdemeanor.

But the happening that Harriet Morrison remembers most vividly concerns the time her husband was accidentally crushed by a horse while working on the ranch. He suffered serious head injuries and was sent to the hospital in Kamloops. The doctors who treated him shook their heads and said they did not think he would be able to walk or see again. They did their best, but after a while they sent him home, saying they could not operate. Martin went to see him twice a day for weeks, sharing attunements with him. Morrison made a full recovery.

24.

A QUANTUM LEAP

Martin's immediate family began to expand in 1980, with the marriage of his daughter Marina to Peter Castonguay.

Born in Ottawa, Castonguay attended Bishop's College School, near Sherbrooke, Quebec. Both his father and grandfather served as chief electoral officer for the federal government for many years. As the hippie era of the 1960s reached its peak Castonguay traveled west to British Columbia and became a long-haired, placard-carrying activist at Simon Fraser University in Vancouver. One day in the fall of 1969—while he was on a break from a picket line—he spotted a poster in a hallway advertising a public meeting with the theme, "Oneness is the Law of Life." The statement agreed with his philosophical thinking, and he decided to attend. He found himself listening to a man only a few years older than himself who, even though he wore a tie and had short hair, was making good sense. At a certain point in his talk the speaker—Michael Cecil—asked a question which touched a deep chord in Castonguay: "While everyone agrees that peace and love are wonderful, who actually has the courage and discipline to live out these ideals?"

The young man from Quebec had explored many avenues of New Age thinking, but he knew that there was a big gap between his theory and what was happening in his life. As soon as the meeting ended he went to Michael. "I want to go to 100 Mile House and see for myself what you're doing," he said, adding, for the sake of a little humor, that perhaps he could ride in the trunk if there was no room in Michael's car.

At the end of his university term Castonguay traveled to 100 Mile House, discovering, like others of his generation, an immediate resonance with what was happening at the Lodge. He worked for a while at a local machine shop and then moved into the mainstream of the Lodge community. During this period he met Marina Cecil, and in the late summer of 1974, after Marina had graduated from Peter Skene Ogden Senior Secondary, they became friends.

A capable, energetic young woman with her father's clear blue eyes, Marina studied art and merchandising at Langara College in Vancouver and attended a course at Sunrise Ranch before returning home to work as a receptionist at Red Coach Inn. Castonguay was now cow boss of the Bridge Creek Estate Ranch. The friendship between the two continued, coming to a culmination—or new beginning—in June 1980, when Martin married them at 100 Mile House. Following the vows and a closing prayer, Martin beamed at his daughter and son-in-law and said, "You may greet each other and then let your shining countenances bless the assembled company." An ice sculpture of two intertwined swans formed the centerpiece for an outdoor reception on Martin's lawn attended by some 275 guests.

At the time of her wedding, Marina was managing a small health food outlet in the Lodge. Drawing on her training at Langara, she gradually expanded the business, until in June 1981 she opened a gift shop in a more favored and accessible location. The "Inn Shoppe"—facing into the foyer of Red Coach Inn—has flourished under her direction ever since.

From her earliest years, Marina's relationship with her father was deep but unobtrusive. "It was a silent relationship in a way," she recalls. "Mum and I did the chattering, and he was just there, looking after his world. But while he didn't have a lot to say in terms of child-rearing, we both knew our connection was solid—it was never disputed as far as I was concerned. And though we often did things together quietly, without a lot of talking, he'd say he loved me. He wasn't *that* quiet!"

Martin and Marina relished jigsaw puzzles, and would work on them together during the long winter evenings. When she was eight, Marina accompanied her parents on a visit to Burghley House, where amongst other new and exciting experiences she spent many happy hours playing with some puppies. When they returned home Martin bought her a puppy, a cross between a cocker-spaniel and a golden retriever that she called "Leo."

In September 1980 another wedding took place. Martin married Luke Vorstermans and Helen Meeker, the youngest of the three Meeker children, whom he and Lillian had raised as their own. A warm, lively woman, Helen grew up with a love of horses and the outdoors, as well as areas like singing and drama. Martin always loved her large heart and urge to exploration. Her husband, originally from the Netherlands, had come to 100 Mile House from Calgary to take a one-month art of living course, and stayed on. He became publisher of the 100 Mile House Free Press, a weekly newspaper.

There was one more wedding to come. In June 1981 Martin united Lloyd Meeker and a slim, self-assured young woman from New York named Paula Martin. Lloyd, born in 1948, was the second child of Lloyd and Kathy Meeker. He attended the University of British Columbia, studying Greek and classical mythology among other subjects, and later received an education of a different kind when he accompanied Bill Bahan on an extended six-month speaking tour through the United States. Sensitive and penetrative, Lloyd has published a book of poems and for a few years conducted a twenty-five piece orchestra. He serves as a faculty member on Emissary classes.

The Castonguays and Vorstermans help coordinate the activities of the Lodge community at 100 Mile House, while the Meekers direct the Emissary group at Denver, Colorado.

Over the years, beginning with the little Piper Colt trainer, Martin had flown in a variety of airplanes, each new machine bringing improved operation and safety. In July 1980, however, the Emissaries made what amounted to a quantum leap in the area of aviation, acquiring a Cessna Conquest propjet. With a cruising speed of 300-400 mph, and an altitude of 35,000 feet, the Conquest replaced a Cessna 421 known as the "Spirit of Sunrise."

Speaking at Sunrise Ranch, Martin touched on the significance of the new plane from a spiritual standpoint. Its ability to fly at a much higher altitude, he said, symbolized what was now required of Emissaries—a shift from the stance of a spiritual student to the stance of authority in living. "Whether anyone has seen it or not," he said, "or whether anyone likes it or not, the new aircraft portrays the fact that it is necessary for whoever is capable, to function in this new space. Our pilot, Charley Kitchens, would tell you that much of the experience that was gained in operating our earlier types of aircraft doesn't fit insofar as the Conquest is concerned. Let us see this with respect to our own experience spiritually speaking."

The purchase of the Conquest was indeed symbolic of a blossoming in the Emissary program. For many years Martin had played the role of a spiritual parent, providing an example of what it means to live in phase with the rhythms and purposes of spirit. Because of this example, and the training he offered, he opened the way for others to know the same experience for themselves. As the process unfolded, the relationship between student and "parent" naturally changed. The awareness developed among all concerned that they were friends, handling what needed to be handled together in a mature and self-reliant way.

The evidence of this coming of age showed dramatically in the summer of 1981 when the Emissaries made a major thrust into the public arena, hosting the 8th annual Human Unity Conference in Vancouver, B.C. Martin did not appear at the event. Instead, representing the increased maturity and authority emerging in many, his son Michael gave the keynote address. Articulating the current of spirit just as his father had learned to do, Michael told the audience, "We can sign off from trying to manage the universe."

I'm sure that we've all come here bearing certain burdens from the past. Perhaps we thought they weren't burdens, but good ideas and valuable experiences. However it is important, I feel, to be willing to come into this place with flexibility, with a real willingness to let life show us what IT has in mind. That might sound simple-minded, but you know, for all the cleverness that has been present in our world, for all the sharp minds that have been getting sharper as the decades have passed, there has been very little wisdom on the basis of mental concern alone. Marilyn Ferguson says in her new book something about the heart being considered as the seat of intelligence in past ages. With all the troubled emotions that have plagued the world, and have no doubt plagued us, that thought may seem unimaginable! Yet it is true, because the only way in which the directions of the compulsion of life may be heard is through the heart, not through our clever minds—at least not first through them.

It has been proved, certainly to my satisfaction and I expect to the satisfaction of many others here, that our minds are very good at understanding after the fact. Sometimes we try to second-guess life, and that is called rationalization. Once the heart is allowed to clarify so that it is no longer like a troubled and disobedient animal, then it begins to be possible to sense with increasing accuracy what life is telling us to do. That is the basis for the restoration of humankind, not because our clever minds come up with new agreements, new arrangements, but because we develop within ourselves the ability to perceive what it is that life is telling us. The supreme intelligence of life, which has brought together the universe and brought together my own body in all its intricate detail, is quite capable of resolving the tiresome problems of men and women who are trying to work out some semblance of order in our present civilization. No problem at all!

Initiated by an Indian spiritual teacher named Sant Kirpal Singh, the Human Unity Conference is held annually in different parts of the world and sponsored by a variety of groups. Approximately nine hundred people from every continent and from about twenty countries took part in the 1981 gathering, held July 23-26 at the University of British Columbia, with George and Joelle Emery as co-presidents. Speakers and presenters included Marilyn Ferguson, author of *The Aquarian Conspiracy;* Jerry Jampolsky, author of *Love Is Letting Go of Fear;* and well-known American poet, Hugh Prather. The program included a women's forum coordinated by Nancy Cecil. Approximately two hundred people shared in this heart-touching event, which launched Nancy into the broader women's arena and led to increasing public speaking with her husband, Michael, and the publishing of her first book, *Magic at Our Hand.*

Following the main program at UBC, a closing session was held at the Queen Elizabeth Theatre, attended by about three thousand people. Parallel events were held in a few other places, notably Boston, Massachusetts, where the mayor pronounced July 25 "Human Unity Day." Thousands gathered on Boston Common to celebrate human unity and hear messages from George Emery, Bill Bahan, Marilyn Ferguson, and Dr. Jampolsky broadcast over a loudspeaker system.

"The creative spirit of God is in action," commented Martin after the Conference. "Something is coming loose, something is coming free, and in the days to come there will be other such indications coming thick and fast." He underlined the need to be able to provide stability and understanding as this process of "coming free" continued.

Because it was the first time Emissaries had been involved in such a large-scale affair, the 1981 HUC was a valuable training opportunity. It emphasized that what was required was not grand concepts or ideas but the ability to connect with people in a sensible way and to affirm their own sensings of a new purpose and direction in life. Humankind is one body. Everyone is included in the action of spirit. An "emissary," in the final analysis, is anyone who participates in that action because of the quality of his or her living.

On September 16, 1981, the Emissary movement celebrated its forty-ninth birthday. On the eve of its Golden Jubilee, Martin announced two more "quantum leaps" that were in the offing. He proposed to make a world tour in February 1982. And plans were afoot to convene a worldwide Emissary congress in Fort Collins, Colorado, in July 1982.

25.

A HALO AROUND THE EARTH

Martin and Lillian, with Grace Van Duzen, left Vancouver in February 1982 on the first leg of a round-the-world tour. In New Zealand and Australia they were accompanied by the Bahans and the Meekers; in South Africa and England, by Michael and Nancy and the Maskells.

As he journeyed through parts of what was once the British Empire—later the Commonwealth—Martin sensed a blending at work in himself. It was a unification of spirit and his own earthly heredity, with its roots going far back to the foundations of the British Empire and the Church of England. This blending included many of the people whom he met during his travels, especially those who were awakening to spiritual identity; and so a spiritual "halo" was woven around the earth.

Martin received a royal welcome wherever he went. At the same time he offered a welcome of his own—an invitation to join him in a place of quietness, assurance, and peace. We are not our bodies, he said. We are not our minds or emotions. We originate in spirit, and have incarnated on earth to let spirit act on the planet.

Such action does not require human forcefulness, he suggested in a talk at Gawler, South Australia. All that is necessary is to let spirit express into the little affairs of daily living. On this basis nothing is insignificant. Others may judge it as insignificant, but it isn't. Martin used an example that occurred when he and his companions were in New Zealand, preparing to fly to Sydney. At the last moment, they learned that they needed visas to enter Australia—something the travel agent had not told them. "You won't be able to continue your flight," an official said.

What to do? From Martin's standpoint it was very simple: continue to bring to bear a right spirit, and what needed to happen would happen. If he needed to be in Sydney, he would be in Sydney. There was no need to try to make something happen. In fact such effort usually complicates things and makes them worse. In the end an airline official phoned the immigration service in Sydney and gained permission for Martin and his party to proceed as planned. "So we cleared immigration in Sydney," Martin related during his Sunday morning presentation at Gawler, "and it all worked out very simply."

But for a while there it looked as though there was a block and that it would really need some manipulation to get through. Let it alone! All that is necessary is to be in place so that the spirit can have the facility for action. The spirit is present: let it act, and whatever way it should come out it will come out. It may not always come out the way the human mind would like to see it come out. However, that's beside the point. Let it be what it ought to be, and we find that all things do in fact work together to perfection when we're in place, when we may say of ourselves, "I came forth from spirit, and am come into the world. There are things to be done here and they are going to be done. The human mind, my human mind in particular, may not know exactly what those things are or how they're going to be done." That's beside the point, because they are going to be done by spirit. But the mind will know what needs to be known in any given situation, as much as needs to be known. Isn't that enough? Do we have to know everything about everything? That would be utter confusion! No, just as much as we need to know, that's all that is necessary. The spirit in action will achieve what needs to be achieved without any sweat of the face.

Gawler is a town north of Adelaide. Paul and Libby Blythe and their group had bought a caravan park there in 1979, and this was now the Emissary headquarters in Australia. Martin spoke to a gathering of fifty-six, his talk later being entitled "Not of the World." In deference to the intense heat, he wore shorts to the meeting—quite proper, he had been assured, in Australia. In view of the heat, and with several other gatherings slated for the succeeding days, someone suggested renting an air conditioner. None of the buildings had one at that point.

"I've noticed people usually opt for the easy answer," Martin commented when he heard the idea. "Why don't we see if we can make something?"

Immediately after lunch everyone, including Martin, scrounged the workshed for suitable material such as eavestroughing, old burlap sacks, and irrigation hose. Using an electric fan, they concocted a makeshift evaporative cooler. "It was only in place for the hour or so that it was needed—long enough to prove a point," Blythe recalls. "But it brought a cool change. I shall never forget the small circle of men—including this elderly British lord—working with hammers and tin snips in this dusty old workshed. Humor mixed with ingenuity as we created something far more than was tangible to us then."

From Gawler Martin and his companions flew to Perth, Western Australia, where they were met by the Maskells, and then on to Mauritius and South Africa, where Michael and Nancy joined them for the rest of the tour. The South African itinerary included a visit to the Hohenort Hotel in Cape Town, and to Emissary centers in Johannesburg and Durban. A happy contingent came to Johannesburg from the Zulu village of Empumalanga to meet the visitors from Canada. A special friendship had been developing with the people of this small African village since Grace Makhathini, wife of Absalom, an elder of the village, came to work at Fourways, the Emissary center in Johannesburg. Martin met the Zulu party at Fourways, the gathering being marked by a great sense of agreement and fun —even though Martin and his party certainly couldn't speak Zulu. As he commented later, "Although we had some interpreters present, the sense of agreement came long before anything was interpreted, because there was a sharing of the same spirit. While this is becoming more or less familiar to us, it was somewhat surprising to the Zulu people who were there." The visitors expressed their delight by breaking into some Zulu songs at the end of the evening.

The friendship and transcendent perspective shared during the tour was poignantly conveyed in a letter which Martin received from a white South African. Martin read the letter during a Sunday morning address at the Hohenort:

I have heard it said that Africa represents the heart, and certainly you sense in the African people the spontaneity and warmth of a childlike innocence that is totally genuine. You also sense the tangled jungle of mistrust and hurt and the wide open plains lying fallow and waiting for a right tiller. The interesting thing is that I am not a black person and yet I know how deeply I am a part of Africa. And there is hurt and mistrust between white and white, as well as white and black, within this heart called Africa.

But pondering further on this, I have felt more than anything what it is to be an Emissary and how my sense and source of identity and home lie in this, not in this continent which I represent. This transcendent kingdom which is my home is shared with others, who each themselves represent an aspect of this diverse continent and more besides. In this lies the essential dynamic for change and healing. Our simple interactions and friendship, based in love for one another and trust and above all a passion for truth, are the foundation and source of all healing. In that sense the world is one kingdom with many parts, and the essential unity lying behind the mind and heart is present, awaiting release through my expression in simple things.

Having read the letter Martin continued: "Each one could say this. It is not, of course, just a matter of Africa, certainly not just southern Africa; it is a matter of the whole world. When we share the transcendent view because we are in that transcendent place and setting, then we stand back of all the events of the world which claim so much of human attention and we begin to see with clear vision what it is that engenders these events." That influence, Martin affirmed, is the spirit of God rising up from beneath the "crust" of human events and human experience. To try to sort out those human events, he suggested, is futile, because the human status quo is in process of dissolving. It is much more profitable to associate with what is happening above and below the "crust" from the standpoint of spirit.

Having touched many hearts on the African continent, Martin, Lillian, and Grace Van Duzen waved good-bye to old friends and new ones at Johannesburg airport and continued on the next leg of their journey, to England.

"It so happens that if there is an institution in Great Britain which is not susceptible to any improvement at all, it is the House of Peers," says a character in Gilbert and Sullivan's "Iolanthe." Undisturbed by trivialities such as age or elections, the House of Lords finds time to take a quiet look at everything. Its approximately 1,180 possible members—ranging from past and present cabinet members to scientists and academics—display a range of interests that is extraordinary. In one year, Lord Ferrers spoke on the following: deer antlers, derelict land, Ethiopia, fruiting plum trees, tinned puddings, whales, widows' pensions, interest rates, oak wilt, Saudi Arabia, drainage, and theatre tickets!

The Upper House of the British Parliament is the oldest legislative body in the world. It conducts its business in an ancient gothic chamber whose stained-glass windows admit a frugal amount of light. The room is furnished with scarlet benches and gold hangings, while portraits of barons from the time of the Magna Carta gaze down from the walls. Speakers employ old-world courtesies. Military men are addressed as "noble and gallant lords," and bishops as "right reverend prelates." Members call each other "my noble friend."

The House of Lords does have some real powers, despite the jokes which are made about it and the unending dialogue about its future. Among other things, it is the highest court of appeal in the land. It can delay the passage of a Commons bill for up to a year, and is the only institution in Britain able to prevent a government from extending its own life without calling an election. Also it does exert a considerable influence over governments, which often will seek to avoid a full-blown confrontation with the Upper House. Financial remuneration is negligible in comparison to that received by United States Senators. It has been said that given London prices, the amount is "about enough to qualify for a Salvation Army hostel."

Martin had succeeded to the title of 7th Marquess of Exeter with the passing of his older brother, David, in October 1981. The 6th Marquess had served with distinction in many areas, having been an active member of both Houses of the British Parliament, Governor of Bermuda, chairman of a major company, and a director of several others. He was also, as has been mentioned, a noted amateur sportsman who served for many years as a senior member of the International Olympic Committee and worked diligently to keep the Olympic Games free of political and other manipulation.

It is a long way from the rolling green meadows and forests of the Cariboo to Westminster, but on March 30, 1982, Martin entered the hushed, red-carpeted corridors of the Upper House to take his seat. Serena Newby, his parliamentary secretary, and the Earl of Lindsay, a member of the House, accompanied him, as well as members of his family. He had invited his old friend, Sir Michael Culme-Seymour, to attend. Unable to do so, Sir Michael sent his best wishes and added, "I hope the House of Lords may one day prove a fruitful avenue to help your work and within which to contribute in your own particular way."

The Falkland Island crisis was coming to a boil by March 30, and the then British Foreign Secretary, Lord Carrington, appeared in the Lords during part of the session, speaking and answering questions about the emergency.

In taking his seat, Martin first of all participated in a swearing-in ceremony, in which he acknowledged his allegiance to the Queen. He was then greeted by the Lord Chancellor, Lord Hailsham, seated on the traditional woolsack; this is a red cushion, stuffed with wool, that recalls the days when Commonwealth trade revolved around wool. Martin's son, Michael, had his own role to play in the proceedings. In accordance with custom, he was required to stand for half-an-hour before the members of the House, giving peers a chance to look over the new member's son and heir.

While he was in England Martin visited Mickleton House, in the Cotswolds, and gave his first address there. Located in the Gloucestershire village of Mickleton, the new property had been bought by U.K. Emissaries in the winter of 1979 as a headquarters for their activities. With the Emissary program still in its early stages in England, Martin used the opportunity to talk a little about that program—what it is, and what it is not.

You can't explain the spirit. It is something to be experienced; otherwise you don't know what it is. You may believe in something—a lot of people have beliefs. By the way, the Emissary program has been looked upon by some as some sort of new religion. It is understandable that it should be seen that way, because even some of those who are associated with the Emissaries have seen it that way. But a religion, after all, if we could define it this way, is a system of worship and belief.

Belief is an indication of the fact that a person doesn't know. If you have to believe in something you don't know it. We don't have to believe in the air we breathe, do we? "I believe in the air!" No, we just breathe it. We have the experience, we know it. We may not be able to describe what the air is. We may not be sufficiently educated to know that it is composed of different gases and what they are and what proportions they are in; most of us don't anyway. But we happily breathe, and because we breathe we stay alive for a while. Knowing may be compared to breathing. We need to breathe the breath of life, breathe the spirit of God in our living, and to the extent that we do that we know it; we don't have to make statements about it; it just is; it is so. On that basis true understanding comes, understanding with respect to all things to the extent that we need to understand them.

Martin's tour concluded at Sunrise Ranch. In his dry way, he commented following his return, "We completed the journey from 100 Mile House to the Ranch in thirty-five days—going the long way around . . ."

Four months later, on the evening of July 29, 1982, Martin stepped to the lectern in the main ballroom of Colorado State University in Fort Collins, Colorado.

"My, oh my! A room full of friends," he exclaimed, as he gazed at the fifteen-hundred men and women who, under the impulse of that spirit which transcends cultures and geographical boundaries, had assembled from all corners of the earth to share the first Emissary International Congress.

This unified body of people had been born out of one man's passion to know the truth in living. For the last thirty or so years Martin had been responsible for its care and unfoldment in the maturing processes. During that time he had seen thousands of people draw close—for a day, a week, or longer—to this growing spiritual entity, and what it represented. He had watched with sadness as a large majority simply went on their way, unaware of what it was they had touched.

Martin knew that this "body" was not yet mature, was not yet a clear or adequate revelation of the one universal spirit that alone brings real happiness and fulfillment. That was one of the reasons he had been delighted when Bill Bahan suggested the idea of an international gathering that would make possible a greater intensification of that spirit through all concerned. But he knew too that those who touched the Emissary program were not being invited to believe some human doctrine or philosophy. They were being shown an open door, not into an enclosure, but into the true freedom of life. Where there was integrity, as had been amply proven now, nothing could stop a person from going through that door and discovering their larger potential.

Just as an individual human form takes shape in the womb under the wise hand of spirit, so had it been with this formation now assembled before him in the Colorado State University ballroom. No human mind—certainly not his—had brought this unique collection of people together. Just as the parts of a physical body are specific, both in their shape and their location, so was it also with this grouping. There was a design in evidence, nothing that he had imposed, but a living design brought forth by spirit. In the measure that all concerned were open to that design, to the compulsion of that spirit, here was an operating, functional organism for the use of spirit—tiny though it was relative to the larger population. This sometimes troubled people who came to associate with him. How could such a small number of people do anything? they wondered. They did not realize that here was simply the tip of a spear—that there were many others, all around the globe, responding to the same spiritual compulsion even though not calling themselves Emissaries.

The Congress dealt with many themes, both spiritual and temporal. One evening, those present went on a "world tour" in which the spirit of different peoples was portrayed in language, dance, customs, and costumes before a hand-painted backdrop of the world. There were also smaller forums on a wide range of subjects—from English as a Second Language to the Right Use of Media—along with music, drama, recreation, and folk dancing.

Martin commented at the close of the gathering:

I think something has happened here during the course of this Congress, something which was symbolized rather remarkably by the fact that when it was suggested you all stand the other evening you all stood, instantly. I don't think that many here present heard the suggestion even. There is a compulsion. When we are all together moving in the current of the spirit something happens very easily and very quickly, something which is most evident because of our proximity in the physical sense. But there is no difference to the experience when we are geographically separated, because the experience comes by reason of the unifying spirit, which is still there just as surely as when we were together in person. Perhaps we can be assured of this, more so, now that it has been experienced by reason of this Congress. It was so before; we didn't really need to come here to have this experience; but seeing that we did come here we might just as well have the experience. And having the experience, we know. Here is a body which remains a body no matter where the parts may be, because the fact of the body is not established in the form as such; it is established in the spirit. As long as we share that spirit we share that body and it doesn't matter where we are. If it is necessary for everybody to stand up, everybody will stand up the world around!

This has seemed a fanciful idea for most people, Martin concluded. "How could this be? We have to figure it all out; we have to have computers and telephones and everything else to get it all coordinated." Such things are useful, the way things are, he agreed; but fundamentally anything that human beings are able to invent is already present in essence in spirit. All that is necessary for intelligent, effective expression on earth is already present.

26.

OF STARS AND SOLAR SYSTEMS

From time to time in his talks Martin touched on the grand picture—the factors at work in our solar system and galaxy, for instance—emphasizing the effects these larger influences have upon the earth, and upon the earth's inhabitants. While humankind has abandoned the idea that the earth is physically at the center of the universe, the human race still tends to think of itself as the apex of intelligent life in the cosmos. Martin considered this an appalling conceit. The wisdom and power apparent in the operating universe, he suggested, dwarfs the intelligence of human beings.

Martin also suggested the possibility that far from having risen from an amoebic slime of some kind, humankind as it is today reflects a process of devolution or degeneration. The early myths and legends of all peoples give support to the view that the earth was once inhabited by a race of men and women who were God-like in their expression and attributes. No confusion, turmoil, or conflict marred that Golden Age because the wisdom and harmony of the universal spirit flowed freely through all without hindrance. The Greek poet Hesiod wrote, "... they lived like gods without sorrow of heart, remote and free from toil and grief: miserable age rested not on them." The Roman poet Ovid said, "The first age was golden. In it faith and righteousness were cherished by men of their own free will without judges and laws ... Earth herself, unburdened and untouched by the hoe and unwounded by the ploughshare, gave all things freely." The Hindu Mahabharata states: "The Krita Yuga (First Age) was so named because there was but one religion, and all men were saintly: therefore they were not required to perform religious ceremonies. Holiness never grew less, and the people did not decrease."

But then came what has been called the fall. In the Taoist scriptures Kwang Tze laments that "the paradisiacal state of the early ages was disturbed by lawmakers. Decadence set in ... and continued until the people became perplexed and disordered." The Hopi Indians tell how people "began to divide and draw away from one another—those of different races and languages, then those who remembered the plan of creation and those who did not. There came among them a handsome one—in the form of a snake with a big head. He led the people still further away from one another and their pristine wisdom. They became suspicious of one another and accused one another wrongfully until they became fierce and warlike and began to fight one another."

The idea of a "fall" has been resisted by many, particularly in more recent times. It is more complimentary, from the human viewpoint, to believe that humankind has been gradually evolving over eons of time into a more ideal and civilized state. And besides, if man actually was a perfect creation, how could he have messed things up so badly? It does not seem reasonable to the human mind.

But suppose—Martin would sometimes submit—the factors involved in this whole situation are of much greater scope than has usually been seen or recognized? Suppose, for example, it is true that humanity is part of a larger picture, a larger, cosmic order and movement, whose very nature is one of change, progression, continuing expansion and creation? Is it not possible that at some point in the long ago, humankind—even though it had been functioning in perfect accord with that larger design until then—simply failed to keep step with some particular change that was occurring in that cosmic harmony? The misstep may not have been entirely deliberate. Even if it was deliberate, the initial divergence may have seemed a small thing—quite harmless. If the only experience that anyone had had to that point was of perfection, how could anyone have been able to foresee the dire effects that would result as that initial misstep was taken—and sustained? It may have been a long time before the terrible consequences were evident: by which time the original tangent may have been so far in the past as to have escaped from memory. One way or another it does seem as if this is the predicament we are in; that we have not only forgotten how or when things went askew, but are reluctant even to admit that they are askew.

The point is not really to look back in any case, of course. Nothing can be done about what happened, or did not happen, back along the way. We have the situation we have now: what are we going to do about it? From Martin's standpoint the single necessity was to acknowledge for oneself the fact of mis-alignment with the universal spirit, the universal whole. Once that acknowledgment is made, so that a person stops trying to cover over the sense of emptiness within, that emptiness can begin to be filled. It is filled as one manifests the true nature and character of love: in no other way.

On a beautiful New England evening in 1983, while visiting the Emissary community at Green Pastures, New Hampshire, Martin sat down and wrote of the crisis facing humankind. He shared what he had written in an address the following morning:

It is difficult not to be aware of the imminent and terminal danger which, like an ominous black cloud, envelopes the whole human race. Many of the more honest and clear-sighted in high places on the world scene realize that virtually anything that they or anyone else may do can at best only delay the inevitable. That inevitable is too massive and too awful even to attempt to describe.

Yet millions attempt to live as though the world continues to be as it has been. Ambition and avarice still prevail. Self-serving and self-indulgence remain the order of the day, and are often equated with freedom.

Do I lay these things before you this morning to engender fear? Fear, among other destructive compulsions, is the cause of what is now coming to culmination. Am I suggesting that it is too late; there is nothing for us to do anymore? You know that the word to be spoken to all with ears to hear is: fear not! Behold, a door is opened in heaven. Come up hither. See and act from this new perspective. The former things are passing away. What value in trying to correct or change them? Let the radiance of spirit be the power of our living, that the new state may be unveiled by the passing of the old.

Everything contained in the old state is now in the past. Is anything gained by analyzing and discussing it? Such action maintains identification with it. Let the former things pass away. Let them go or go with them! No one is exempt from the working of the creative cycle.

What is the spirit to be expressed now? What is the attitude to be taken in this moment? Always for us these spring out of passion for the truth, never from reactive judgment of people and circumstances.

It may be said that the day of decision is here. This is the day; this is the hour; this is the moment. Only in calm and understanding tranquility can the end of darkness come and the life, which is the light of men, rise with healing in his wings.

In July 1983 Martin received news of the death of Dr. Bill Bahan, stricken with a heart attack while en route to a speaking engagement. Bahan had played a vital role with him, bringing encouragement and inspiration to many thousands of people and above all maintaining a clear, victorious stance in his life. In a thanksgiving service Jim Wellemeyer, speaking at Green Pastures, said it was significant that people from all around the world were gathering in tribute. "Bill was a world traveler," said Jim. "I think this globe, which we have to my left, is symbolic of something as it relates to Bill. Someone in a note said that when Bill was present the world lit up. And how true that is. He brought the light." Speaking from 100 Mile House, Martin said that Bahan, whom he had known for about twenty-four years, had quickly become his loyal and trusted friend. "We shared vision and understanding in a mutual recognition of true purpose, which has nothing to do with the intellectual theories of men, even if those theories are called spiritual," said Martin. "It is a matter of being naked and unashamed before God, and then there is the beginning of wisdom. Bill was wise and a friend to everyone whom he ever met."

Martin celebrated his 75th birthday at 100 Mile House the following year. Increasing numbers of people were now visiting the Emissary communities at Sunrise Ranch and 100 Mile House, drawn by the strong, steady pulse of spiritual expression emanating from the two places. "Much has changed here over the past forty years," wrote Roger de Winton in the Sunrise Ranch magazine, *Eden Valley News*. "The need to host people of varied backgrounds and customs from all parts of the world with intelligence and understanding grows with the passing years. Yet all of this is done with ease as the invisible influence back of all is allowed to emerge in its expansive way and in the joy of creative giving. So the Sunrise community continues in healthy, vibrant service."

It was the same at the Lodge community at 100 Mile House, where Martin's son, Michael, wrote in *Northern Light*:

In the building where we meet hangs a large map of the world, mounted on canvas and framed in oak. In a way it represents the consciousness that exists here of wholeness. Our community, located in 100 Mile House, three hundred miles north of Vancouver and far away from major happenings on the globe, has a special opportunity for the maintenance of an uninvolved perspective.

Northern Light is an example of what is actually being done in this regard. The simple right handling of ordinary tasks and affairs as spoken about in the majority of the articles is seen as immensely important. Indeed, the proper attention to each intimate detail of living is recognized as proof of integrity, and through the nurturing of this spirit in action, oneness is an inevitable result. Clearly the world as a whole is in great need of such an example. The atmosphere of safety and creativity that has been developing here is an invitation to everyone—not to come and live at 100 Mile House but to experience the same potential for themselves where they are.

People from every corner of the earth are our close friends and associates. Throughout North America are many, of course, but also from India to Japan, South Africa to Argentina, Britain to Israel, Australia to Nigeria, and so on. No doubt there are many more unknown than known. The fundamental urge toward oneness, based in the expression of personal integrity, is very widespread. We salute all such as colleagues and would say, through our living, that oneness is genuinely possible, indeed inevitable!

The fruits of the living of the man now known as Martin Exeter were becoming increasingly evident. At the annual Emissary conference at Sunrise Ranch in 1984 plans were laid for an ambitious public event to be held the following year under the title, "The Rising Tide of Change." It was proposed to hold a series of one-day public presentations which would follow one another consecutively, moving from east to west around the world.

27.

THE RISING TIDE

Shakespeare knew something about life. "There is a tide in the affairs of men," he wrote, "which taken at the flood, leads on to fortune; omitted, all the voyage of their life is bound in shallows and in miseries." I wonder how many people have pondered those poignant words. They are surely true. Subtle rhythms and compulsions, arising from their source in the one true spirit, flow like rivers of gold in the lives of us all, and we either move with them or are left "bound in shallows and miseries."

From Martin's standpoint, the Rising Tide of Change events—public symposia hosted by Emissary groups in nineteen locations around the world in 1985—served a twofold purpose. They required Emissaries everywhere to put themselves on the spot, and express the spirit of love, truth, and life openly. They also, and more importantly, drew attention to those causal rhythms that are back of all the surface changes happening on earth. Here was an opportunity for people to get the feel of moving with spirit rather than struggling against it—a painful business, as Martin knew from his own past experience. True, the Rising Tide of Change events included a consideration of such areas as business, education, art, and health. But this was not the real point. As Martin commented in a preparatory talk, "If people come to these events most easily because they are interested in business, for instance, that's fine. But they don't really come to learn more about business. They come to experience what the creative expression of spirit is in living. Of course if they carry that into their business, the business will be transformed and probably somewhere along the line will cease to be what is usually thought of as business."

Sponsored by the Society of Emissaries, the series of presentations began at Los Angeles and Vancouver on April 13, 1985. The events—each one different—moved west to Tokyo, Auckland, and Melbourne on April 25; to New Delhi on May 11; and to Johannesburg, Lagos, and Tel Aviv on May 25. The "Tide" reached London on June 8, then crossed the Atlantic to North and South America, concluding with gatherings at Calgary, Alberta, and Fort Collins, Colorado.

After hearing a report on one event by telephone Martin commented:

We sensed the substance of your time together in this Rising Tide event, which did indeed extend further than your immediate vicinity, giving an awareness of the vast extent and power of what it is that is really moving. Your emphasis upon the spirit of truth, the comforter, is aligned with our consideration here this morning. This expression of the truth has seemed to be out of reach to human nature. The churches have been inclined to teach that it IS out of reach. It is out of reach for human nature, that's a sure thing; but it is immediately within reach, at hand, when human nature begins to be relinquished. And here is the truth, the spirit of truth emerging into expression individually. Let us never think that there is safety in numbers, so that we are always looking around to see, "Well, who else is going to allow this to happen?" If you take that sort of attitude, that's the human nature attitude, and the spirit of truth has no means of expression through such a person. It is an absolutely individual matter. I must do it. I am the one. I am the way, the truth, and the life.

For my own part, the Rising Tide day in New Delhi proved as rich and spicy as tandoori chicken. Joy and I arrived a few days before to help A.B. Singh and others prepare for the event. We were accompanied by Mary Ellen Roos, a singer and composer who wrote the music for the "Rising Tide theme," and Achal Bedi. The room was well filled despite an outbreak of terrorist bombings in New Delhi the previous evening. We held forums on a wide range of subjects and enjoyed some unique artistic offerings. New Delhi's top choir sang the musical invocation, and Roshan Seth, who played the part of Nehru in the movie "Gandhi," read a poem called "The New Nation." Suresh Kulkarni led the group in chant and simple yoga exercises, and the audience sang the Rising Tide theme with gusto and panache after a lesson from Roos. As Joy reported in *Newslight*, "That evening a very hot Indian sun set on those hundred people who had experienced the oneness of spirit which pulses beneath culture, beneath politics, beneath physical circumstances, beneath time and space."

The Rising Tide day in London, England, attended by people from four continents and fourteen countries, included a presentation from several experts on near-death experiences. "It's as if the world itself is in the process of having a near-death crisis," declared author Margot Grey. A popular presenter was Frances Horn, author of *I Want One Thing,* who had been a guest speaker at the Rising Tide events in Johannesburg, Melbourne, and Vancouver. Frances, nearly 80, announced that after corresponding with Martin she had decided to entitle her next book, "Wanting May Kill You." During one open session an elderly lady said, "It's just like coming home." "Life can be lived very passionately but still be balanced," said a woman from Norway.

Martin's son-in-law, Peter Castonguay, was the keynote speaker at the Rising Tide day at San Miguel de Tucuman, in Argentina. The Emissaries had been represented in Argentina and other South American countries since the early 1970s, when Rena Cassidy translated the original *Integrity* letter into Spanish and sent it to various South American newspapers. The letter evoked a particular interest in the "garden city" of Tucuman, where a buoyant group developed in spite of the strife and tension current in Argentina at that time (one present member lost a brother and sister when they were taken away by police, never to return).

Radio and television interviews, and a Decree of Provincial Interest signed by the Governor of Tucuman, welcomed the Rising Tide gathering in the ballroom of Tucuman's venerable Jockey Club. Ana Maria Uruena, one of the first Tucuman residents to reply to Rena Cassidy's letter, spoke some opening words. Diana Soto, of Sunrise Ranch, and

Eric Crocker, formerly of Glen Ivy, California, who now heads up the Emissary group in Tucuman, played a key role together with Castonguay.

In September 1985—a few days after celebrating their thirty-first wedding anniversary—Martin and Lillian flew to Toronto to begin a tour of the eastern United States and Canada. After sailing in Toronto harbor on Lake Ontario, touring one of Toronto's busiest firehalls, and attending a performance of "Cats," Martin spoke in his old stamping ground, the King Edward Hotel in Toronto, to three hundred and fifty people from as far away as Quebec, Michigan, and New York state. "The fact of the matter is that significance already exists," Martin reminded his listeners. "All that is necessary is to unveil it, to reveal it, to let it be brought forth; and it requires us and others for this to happen."

Having reached his mid-seventies, Martin had been in the business of revealing his own significance for a long time now. As he himself aged physically, he sought to meet and dispel the commonly held belief that people lose value as they grow older. On the contrary, he affirmed, an older person rightly has more to offer—and a greater spiritual responsibility—than ever before.

Value is in spiritual expression, not in what you can do physically. What you can do physically may determine somewhat the nature of your spiritual expression, but young people in a general sense have less spiritual substance available for spiritual expression than older people. This is why it was indicated, in the story about Jesus changing the water into wine, that the good wine was left to the end of the feast, because there is rightly far more substance after age sixty-five than before that age. But as everyone has the habit of placing value upon externals and what one may do in a physical sense, it seems as though most of the value is in youth and it is all lost when you get older, which is absolute nonsense. Let us never write ourselves off, and let us never write other people off simply on the basis of the fact that they have spent a few more years on earth than we have. In considering the matter from my own personal standpoint, there are many things that I cannot do now that I used to do when I was younger, but I certainly do not take the attitude that because of that I am becoming more and more useless. This unfoldment of years has provided space for the increasing generation of substance through which the greater intensity of spiritual expression may emerge. We all need to be on hand as long as we are needed to be on hand, and really on hand, spiritually speaking, with the substance and the ability to handle the creative action that is necessary. From my standpoint the older people have far more responsibility in this than the younger people.

At one point in the development of the Lodge community, a person who had been given a certain area of responsibility became a target of dissatisfaction. There was a feeling, on the part of some, that he was overstepping his authority and behaving in an arbitrary manner. Eventually a meeting was arranged to discuss the issue. As various ones expounded about all the things they felt were wrong with this individual's approach, you could feel a strong current of condemnation building in the room. The complaints all seemed so reasonable, after all; so justified . . . until with a few succinct words Martin cut through the criticism.

"Surely we are not here to blame or criticize anyone," he said quietly. "We are here to love one another and support one another because we honor the true nature of love." His words pricked the heavy atmosphere as if it had been a balloon.

The attitude of blame has bedeviled humanity for as long as memory goes back. It is considered so normal and natural that it rarely excites comment. And yet it is a primary cause of all the divisiveness and suffering on this planet. It is the block that prevents the natural experience of moving with spirit, of *being* the truth of love in expression.

Because he met this perverse human trait of blame clearly and unequivocally over a prolonged period Martin pioneered a new human state—a new norm—as different to the old state, the old norm, as day is to night. This new state, characterized by the natural flow of respect, thankfulness, and appreciation, must always begin with the individual. Increasingly, however, inspired by Martin's example and vision, groupings of individuals emerged in various parts of the world revealing the true character of love collectively.

28.

FINISHING THE WORK

The Haram, the massive oblong courtyard of the Great Mosque at Mecca, is large enough to hold half a million people. Arches and colonnades of white marble rise two stories high on all sides. Entering the Gate of Peace that leads into the courtyard, pilgrims who moments before were shouting praise to Allah fall silent. The Haram was a sanctuary even before Islam; it is said that the birds that flock to Mecca refrain from flying over the Great Mosque so as not to disturb the peace and harmony below. In the center of the courtyard stands the Kaaba, a stone structure built by the patriarch Ibrahim over three thousand years ago as a house of worship to the one God. "The Kaaba is very, very simple—little more, really, than a huge cube with a single door set six feet from the ground," says Jehan Sadat in her book, *A Woman of Egypt*. "But the power of this building is boundless. It is toward the Kaaba that Muslims everywhere, one fifth of the world's population, turn five times daily and bow down in prayer."

While the Great Mosque occupies a position of unique importance for Muslims, there are many other sacred sites and buildings on the face of the earth, places where people may renew a sense of connection with a higher source. Entering a building such as Westminster Abbey, in London, or some of the ancient Hindu temples of India, one is keenly aware of a quality of atmosphere that seems to exude from the very walls themselves. And yet I have often thought that if there is a sense of peace and sacredness in, let us say, the Great Mosque of Mecca, or Westminster Abbey, or a small, white-painted church in Massachusetts, the peace and sacredness come, ultimately, from people. People built these structures in the first place—built them with a deep sense of love and devotion. And those who come to worship also, for the most part, come in love and devotion. Here is the source of the atmosphere of love and sacredness that may imbue such shrines.

Men and women are the true temple of God—their physical forms created for the sole purpose of accommodating and revealing love and truth on earth. "The greatest city is the city of the greatest men and women," Walt Whitman declared.

"The Lord showed me, so that I did see clearly, that he did not dwell in these temples which men had commanded and set up, but in people's hearts," wrote George Fox, founder of the Society of Friends.

When spirit is accommodated by human beings in this way, then there is a place where all the other kingdoms of the earth—the animal, mineral, and vegetable kingdoms—may find rest and ease.

As men and women awaken to the immense spiritual responsibility which is theirs—and not just for an hour or two on a holy day—some may elect to live in physical proximity with others who are doing likewise. This is a personal matter. Most of those who share in the Emissary program, for example, live on their own, either as individuals or as nuclear families. Others find it natural to associate in a larger community. Consider the human body as an analogy: some cells are included in specific "communal" groupings—the various organs—while others participate in a freer, less structured manner. As we move in the way of spirit there is room for great variety in this regard. There is also an invisible design involved, so that as we continue in this way we see where we should be. Again, this is certainly the case in relation to the physical body. There is a proper place for each cell and part. Heart cells inhabit the heart; the hand joins the wrist; ears extend where ears are supposed to extend.

This all correlates with the emergence in recent decades of groupings of men and women, within the Emissary movement and elsewhere, whose sole concern is to let the light of spirit burn in the earth. Such people bring stability and peace into the world because they express these essences in their living. Here is the emergence of a new culture born of the simple expression of love, love for one another and love for the earth. This new culture does not emerge because of wishful thinking, or because someone gets a bright idea in his or her head and tries to make it work. It emerges because finally there are those with the courage and common sense to relinquish the ways of the old culture—blame and accusation, for example—with their bitter fruits of divisiveness and pain. As this is done there is space in human hearts and minds for light. That is the name of an Emissary community in Israel: "Mishkan Or"—"Where Dwells the Light."

I would like to mention twelve global communities which form part of Martin Exeter's legacy to humankind. They have evolved and matured to their present state because of his example of the spiritual life. Each of these communities is distinctive; each, certainly, is busy; but a quality of stillness characterizes them all as they interact in lively and original ways with society at large. They host seminars, classes, and meetings of various kinds, and are active in such areas as drama, music, business, agriculture, and education. Most importantly, their doors are always open. In relation to visitors, the concern is fundamentally to offer the opportunity to sense and experience—through hands-on participation, if possible—the ease and satisfaction of moving with life's creative flow.

For those who inhabit them, of course, such communities are also intense proving grounds requiring that a clear expression of love and truth be sustained regardless of what is happening within a person, or round about.

Sunrise Ranch and the 100 Mile Lodge community have already been introduced, and also Green Pastures, founded in 1963. These three communities were followed by King View Farm, in Ontario (1972); by the Edenvale community, in British Columbia, and Oakwood Farm, in Indiana (1973); by Still Meadow Farm, in Oregon (1976); and by Glen Ivy, in California (1977). Later four more sister communities emerged, all of them off the North American continent. These were the Hohenort, in Cape Town, South Africa

(1978); Hillier Park, in Gawler, South Australia (1979); Mickleton House, in Mickleton, England (1980); and La Vigne, which nestles in the French Alps in the village of Velanne La Sauge (1986).

Martin and Lillian and other members of their family visited La Vigne in the fall of 1986 after spending a few days at Mickleton, where they enjoyed a cream tea on the lawn of Mickleton House in unaccustomed English sunshine and met a number of British Isles Emissaries for the first time. "I was very impressed by a quality in him of not standing for any nonsense, of being full to the brim with common sense," said one. Another commented, "Somebody asked him a question that seemed aggressive and I felt protective towards him, but he didn't need that. The humility of his reply was beautiful." While in England Martin and Lillian visited Windsor and saw their grandson, Anthony, installed in the 500-year-old halls of Eton College. Anthony, who now carries the title Lord Burghley, is heir to the Exeter title. Born on August 9, 1970, he is an intense, adventurous young man who is equally at home discussing global economics or climbing a mountain. After finishing at Eton in 1988 that is, in fact, what he did. He joined a Soviet-American Outward Bound exchange program that took him to the peak of Mt. Elbrus, highest mountain in Europe. Anthony has a younger sister, Angela, who was born in 1975.

La Vigne—"The Vine"—is an unusual home, covered with thick Virginia creeper. The main building was once a barn used for breeding silkworms. Located in the foothills near Grenoble, the property is an hour-and-a-half drive from Lyons and also Geneva. But La Vigne is neither property nor building: it is the growing family of people from various parts of Europe who live there, and who infuse the estate with their own exuberant love for life. The community was begun by two young couples—Serge and Betsy Nicolet, from Grenoble, whose family had owned the property; and Marco and Sarah-Jane Menato, from Padua, Italy.

Representatives from eight countries in western Europe joined Martin and Lillian and also Michael and Nancy for a five-day gathering at La Vigne, ending on the Emissaries' fifty-fourth birthday. Commented Robert Marcus, of Munich, West Germany, "This was a time of beginning, of newness, and the spirit of victory predominated at the core of our experience with an almost tangible vitality. At last, after all these centuries of being stuck in the ruts of the encrusted human state, the Old World received the New." During this same month of September, the members of the European Economic Community were also meeting, taking more steps in a process of agreement and unification which continues to bring great change to Europe.

Martin dedicated La Vigne during a Sunday morning address on September 14, after spending two hours with a group of men building an improvised lectern and podium. During his talk he used the analogy of a magnet to portray what was happening, in a spiritual sense, in Europe and elsewhere. Just as the lines of force from a magnet draw iron filings into a particular shape, in the same fashion, he pointed out, the power of love draws those who are open to it into the living design of spirit. La Vigne itself had emerged because of this magnetic force.

There is something without precedent occurring, and we begin to have a vision of this, an awareness of something vast. It seems to take a little while before human minds and hearts have enough space to realize the largeness of what is happening, so it has been coming in increments, one might say, a little at a time. Something emerges in this location and that location, with a few people; it works back and forth. But gradually the magnet puts in an appearance and there is the way by which all who will are drawn along the lines of force. This is a different kind of leadership, isn't it? There is no one there cracking a whip and saying, "You get in line." It just happens of its own nature. It happens because of the nature of the people concerned, because of what is inside.

We have pictures in the religious field of some authoritarian god sitting somewhere and judging people. They all come before him and he says, "You did good and you did bad, and here is your reward." That is a very human way of looking at things, isn't it? If you do a good job in your factory you will be promoted; you get a reward. If it's a bad job you get fired. This is all a human concept. The idea has been plastered onto God, whoever God is, somewhere. But God is right here, coming forth through people, those who are willing to let it happen. Because of that there is a magnet and there are the lines of force and there are those special ones who come along the lines of force.

The five-day gathering took place in bright sunshine, with outdoor meals in the shade of four fine lime trees. The program included swimming in a local lake, and a visit to the famous lakeside resort city of Annecy.

From La Vigne the Exeters journeyed to Paris and then to Sweden to visit the land of Lillian's ancestors. Later they spent a weekend at Burghley House with Michael and Nancy and the Castonguays. As they entered the grounds of Burghley, having driven north from London, the Exeters saw the flag of the Marquess flying above the House; by custom, the flag is raised when he is in residence. At Martin's invitation, two generations of the Cecil family met for lunch, the first time that they had all been present at a gathering with their transatlantic relatives. Lady Victoria Leatham, Martin's niece, who lives at Burghley House, took everyone on a tour of the staterooms, the Great Hall, the Orange Court, and other areas which are no longer occupied by the family, but are open to the public. Later the family enjoyed a slide show of 100 Mile House presented by Peter Castonguay. As Lillian reported in *Newslight,* "The halls and rooms of Burghley were filled with joy and laughter and genuine pleasure to be together."

Not long after Martin's return to North America Conrad O'Brien-ffrench passed away at Sunrise Ranch at the age of 92. At a memorial service in November 1986 Martin suggested that perhaps the one whom he and others had known as Conrad had not really gone anywhere. All that had happened was that the limitations and restrictions of the existing human state had passed away. Why—Martin asked—should we wait for what is called death in order to have that experience? Why not allow those limitations and restrictions to pass away now, while one is still alive? He expressed his thankfulness for his long association with O'Brien-ffrench, and for the latter's contribution to the emergence of a new, transcendent state on earth.

Being Where You Are and *On Eagle's Wings*—paperback collections of some of Martin's talks—had been published in 1974 and 1977 respectively. These books were followed in 1986 by another compilation, *Beyond Belief—Insights to the Way It Is.* Reviewing the new work, author Richard Heinberg, of Sunrise Ranch, wrote, "Surely if we human beings are going to reverse the present world trends toward environmental disaster and thermonuclear suicide we must eventually personally address the opinionated, manipulative frame of mind which has created these problems ... It is refreshing to read such a clear diagnosis and prescription in nonsectarian language, and to know that there are men and women from many walks of life who are already experiencing the transformation described in this book, and who are exerting an influence in the world at large."

In April 1987 Emissary Foundation International—EFI—sponsored "The Signs of the Times," a public event which took place in seventy locations in twenty-three countries around the world. An estimated 4600 people considered both the signs of collapse evident in the world, and the signs of creative achievement and transformation that are also apparent.

EFI is the main out-front face of the Emissary program. However, several other organizations have burgeoned in recent years, helping to convey the practicality of spirit into various fields of human endeavor. These include the Whole Health Institute, founded by Dr. Bill Bahan; Renaissance Business Associates; Renaissance Educational Associates; the Association for Responsible Communication; and the Stewardship Community, which is concerned with right stewardship and care of the earth. Each of these associations is autonomous, with its own board of directors. Each has its own newsletter. Each develops its own programs and activities. The Whole Health Institute, for example, holds an annual conference that attracts leaders from a wide range of health-related fields. The sixteenth such conference, held in Estes Park, Colorado, in 1988, considered the radical premise that if health remains elusive, it is principally because we interfere with its intent to happen. The conference sought to expand the experience of being at the core of healing, the place where there is no interference. Here, it was suggested, is the most exciting and challenging frontier of healing, where interest is in the cause of health—not disease.

An example of the impact which these organizations are having in today's world occurred during an "Entertainment Summit" held in the United States toward the close of 1987. The Association for Responsible Communication, a network of media professionals which helped sponsor the "Summit," invited a group of Soviet and American filmmakers to share in a workshop at Glen Ivy; the participants included Elem Klimov, president of the Soviet Filmmakers' Union, actress Ludmilla Chursina, and TV journalist Vladimir Posner.

After a useful and productive day which included some time in the Glen Ivy hot tubs and spa, about sixty people sat down to their evening meal. Martin, as the host, was going to say grace. He wanted his words to be acceptable and relevant, and prior to the meal he talked for a few moments with Bruce Allyn, a member of Harvard University's Negotiation Project, who has been active in various Soviet-U.S. peace initiatives. Martin asked Allyn if there was a Russian phrase that he could use to close his blessing. Allyn suggested the words "cheestaya nyeba," which, literally translated, mean "clear sky." In fact, they mean more than that, perhaps the equivalent of "open heaven." Martin evoked this image in his blessing. As he finished, all present broke into a spontaneous round of applause—a moving indication of how many in our world long for a clear sky and "open heaven."

1987 was the last year of Martin Exeter's life. When he passed away on January 12, 1988, after a brief illness, I thought to myself at first, "This was sudden." With a little hindsight, that idea soon passed. I saw how perfectly the cycles of his life had worked out, how well he finished what he came here to do. A man who would not go to bed until he had finished fixing a living room lamp would certainly not leave this earthly scene until he had done what he needed to do!

With this same advantage of hindsight I saw how during this final year Martin put the finishing touches to his life's work, making sure that the body of people which he had brought together was ready for the task that confronted it. For thirty-three years—since August 1954—he had been directly responsible for the care and development of this grouping, offering through his words and his living a continuing reminder of the noble quality and character of life. This current of inspiration and reminder continued unabated during 1987, though he acknowledged, too, that fundamentally this developing organism was in place—in position to play its part as a vehicle for the release of spirit. His job had been done, in other words. As he said in an address at Glen Ivy in March 1987—"Finally the spirit is being brought to focus in a body of people who can be trusted, not a body of people all sitting in one place but spread around the world. It's one body just the same, because there is one spirit."

Bringing these people through the transforming fires of love to this point of readiness was one essential task which he had accepted, and carried out. Another vital aspect was ensuring the provision of wise and understanding leadership within this growing entity, so that it would be capable of handling itself effectively when he was no longer on hand. Each one rightly shares that responsibility of leadership. But for many years he trained and prepared his own son, Michael, to provide the overall coordination which is necessary. I don't recall that an election was held to decide this matter (although Michael's appointment was considered, and approved, by the Executive Council). On the other hand, if an election had been held, I am sure that Michael would have won unanimously. One knows that ultimately he cares for nothing but the truth in any situation, but he is never arbitrary; like his father, he never tries to tell anyone what to do—because direction, after all, is contained in spirit, in the truth of each person. Rather, through his own serenity, his own clearcut assurance in life's wisdom and power, Michael simply inspires serenity and trust in others. It was because of his pervasive, harmonious presence throughout the Emissary program over a long period of time, and the trust which others knew in him, that the transition when Martin departed could be as smooth as it was.

Symbolic, perhaps, of the shift which was already occurring, Michael and Nancy left Vancouver in mid-November 1987 to bring the Emissary presence, and message, to Japan. During a busy four days in the Tokyo area, they gave two public talks and met with a wide range of people, including followers of a man named Mokichi Okada, whose life, influencing many thousands of people, had paralleled Lloyd Meeker's in remarkable ways. During a visit to a sanctuary at Hakone Michael and his party shared "johrei," a form of healing taught by Okada, in which radiation is extended through an outstretched hand. Michael described later how the minister faced a shrine and chanted, then turned and raised his hand in blessing. "It was a beautiful experience. The radiant current was moving from him to us. It was coming from his own place of inner stillness. Here was a true healer, because he had found resonance with the authentic light, and this overflowed through him."

From busy Tokyo the travelers continued to Australia and

New Zealand. In Gawler, South Australia, Michael commented that the role of Emissaries is "simply to be." There is nowhere to go, he affirmed, nowhere to come from, nothing to grasp, and nothing to convince anyone of. "In this hectic world where all the indicators are going straight down, what could be more helpful than the presence of people who aren't affected by what is occurring, or filled with apprehension, but are content simply to reveal the authentic light of spirit?"

As if he had merely waited for Michael and Nancy to complete their ground-breaking trip, Martin gave his last address on the evening of December 13, 1987, at 100 Mile House. It was later entitled, "Radiant Light: The Transforming Power." Michael shared in the presentation with him, as had been usual in recent years. By the following weekend, Martin was ill with what was at first thought to be flu, but later was diagnosed as spinal meningitis. His work on earth was indeed finished. The light he had brought would shine through other eyes, other minds, other hearts; through an increasingly unified and expansive body. It was all right, now, to leave.

As a cellist, when he has finished playing, lays his cello down, so Martin laid down the instrument which had made it possible for him to make music on earth. He could not have done this, he could not have brought the music of spirit into the planet, without that instrument; yet it did not actually give him life—nor, when the time came to set his physical form aside, was his life in any way diminished. As it is said in the beautiful words of the Bhagavad-Gita, "Never the spirit was born; the spirit shall cease to be never . . . birthless and deathless and changeless remaineth the spirit forever."

29.

OUR TRUE ANCESTRY

What great destiny could this little patch
 Of Cariboo earth hold,
These few tumbledown buildings
Gently moldering away on a lonely road
In a lonely land
Far from the pulse of cities
And the wealth and power of England?
No destiny here,
Common sense would have said
As it surveyed the forlorn scene
Other than the age-old fate
Etched deep in the human heart
—To labor, to dream, and to die.
But he beat the system,
This man I knew as Martin.

He did not allow himself to get away with things.
He did not allow himself to quit.
He kept his integrity.
And as he held fast to that golden cord
The sweet voice of spirit spoke from the stillness
And reminded him of a different ancestry
A different home
Greater than the proud name of England
Or the proud halls of Burghley.
He remembered his place in the serene order
Of the stars,
And in the sunlit courtyards of eternity.
He remembered that he was not just of the earth
Even though he lived here for the moment;
He carried the light of the sun in his eyes.
He was a prince
From the realm of light
And he had come to remind others
—Trapped in bonds of forgetfulness and despair—
That they too were of light
And together their work was to return the planet
To a place of beauty and love.